AF411423

Advances in Selective Flotation and Leaching Process in Metallurgy

Advances in Selective Flotation and Leaching Process in Metallurgy

Editor

Ilhwan Park

MDPI • Basel • Beijing • Wuhan • Barcelona • Belgrade • Manchester • Tokyo • Cluj • Tianjin

Editor
Ilhwan Park
Hokkaido University
Japan

Editorial Office
MDPI
St. Alban-Anlage 66
4052 Basel, Switzerland

This is a reprint of articles from the Special Issue published online in the open access journal *Metals* (ISSN 2075-4701) (available at: https://www.mdpi.com/journal/metals/special_issues/flotation_leaching_metallurgy).

For citation purposes, cite each article independently as indicated on the article page online and as indicated below:

LastName, A.A.; LastName, B.B.; LastName, C.C. Article Title. *Journal Name* **Year**, *Volume Number*, Page Range.

ISBN 978-3-0365-2980-6 (Hbk)
ISBN 978-3-0365-2981-3 (PDF)

Contents

About the Editor

Ilhwan Park received his Ph.D. from Hokkaido University in 2019, and afterward, he was appointed assistant professor at the Division of Sustainable Resources Engineering, Faculty of Engineering, Hokkaido University, Japan. The subjects of his research are the recovery of critical metals/minerals from primary and secondary resources by mineral processing (e.g., flotation, magnetic separation, gravity separation, etc.) and hydrometallurgy (e.g., leaching, solvent extraction, cementation, etc.). In addition, Dr. Park has been engaged in the development of sustainable techniques for the prevention/control of acid mine/rock drainage (AMD/ARD) and the remediation of contaminated soils and water. From his work, he has published papers in leading international journals in the fields of mineral processing, extractive metallurgy, and environmental science.

 *metals*

MDPI

Editorial

Advances in Selective Flotation and Leaching Process in Metallurgy

Ilhwan Park

Division of Sustainable Resources Engineering, Faculty of Engineering, Hokkaido University, Sapporo 060-8628, Japan; i-park@eng.hokudai.ac.jp; Tel.: +81-11-706-6315

Citation: Park, I. Advances in Selective Flotation and Leaching Process in Metallurgy. *Metals* **2022**, *12*, 144. https://doi.org/10.3390/met12010144

Received: 11 January 2022
Accepted: 11 January 2022
Published: 12 January 2022

Publisher's Note: MDPI stays neutral with regard to jurisdictional claims in published maps and institutional affiliations.

1. Introduction and Scope

Metals are a finite resource that are necessary to maintain living standards in modern society, due to their countless applications, such as transportation vehicles, building and construction, household appliances, electronic devices, etc. In addition, the world is rapidly transitioning to "low-carbon technologies" using renewable energy sources (e.g., solar, wind, etc.) to combat climate change, and these technologies require vast amounts of metals per unit generation compared to that of conventional fossil generation; for example, 11–40 times more copper (Cu) for solar photovoltaic (PV) systems and 6–14 times more iron (Fe) for wind power stations [1].

Unfortunately, easily exploitable ore deposits are hard to find, which makes it unavoidable that mining industries must develop complicated ore deposits with low-grade and fine grain-size. The complexity of these ore bodies requires fine grinding to achieve the appropriate liberation of valuable minerals. Typically, there are two scenarios to extract metals from finely-ground ores; that is, (1) concentration of valuable minerals via "flotation" followed by smelting process and (2) direct extraction of metals from finely-ground ores by "leaching" followed by purification and recovery processes (e.g., solvent extraction and electrowinning (SX-EW)). Therefore, flotation and leaching, both of which are the first stage of each scenario, of finely-ground ores are of crucial importance to assure the continued supply of metals. Thus, this Special Issue introduces the latest scientific advances in selective flotation and leaching processes essential for the production of metals in a sustainable manner.

2. Contributions

Eleven articles have been published in the present Special Issue of *Metals*, encompassing the fields of flotation and hydrometallurgy. The papers are all of highly scientific value and will be of great interest to readers of *Metals*. The contents of the published papers will be briefly summarized as follows.

2.1. Flotation

Copper has been widely used in various applications, due to its excellent electrical/thermal conductivity, high corrosion resistance, etc. [2], approximately 60% of which is produced from porphyry copper deposits (PCDs) [3]. The beneficiation of porphyry copper ores is mostly achieved by a two-step flotation process; that is, (i) bulk flotation to recover Cu and molybdenum (Mo) minerals from gangue minerals and (ii) selective flotation of Mo minerals from Cu-Mo bulk concentrates using sodium hydrosulfide (NaHS) as a Cu depressant. Although this process is efficient and well established, the use of NaHS for Cu/Mo separation has several drawbacks. A review paper by Park et al. [3] introduced recent depression techniques, including alternative inorganic/organic depressants as well as oxidation treatments involving the use of ozone (O_3), plasma, hydrogen peroxide (H_2O_2), and electrolysis. Moreover, Park et al. [4,5] developed a new depression technique (i.e., microencapsulation using ferrous and phosphate ions), which preferentially coated chalcopyrite ($CuFeS_2$) with ferric phosphate ($FePO_4$) layers rather than molybdenite (MoS_2). As a result, the floatability of chalcopyrite was selectively reduced, while molybdenite floated well by the addition of kerosene (Mo collector).

As high-grade and easily exploitable copper ores have been getting depleted, the development of flotation techniques for copper ores with low-grade and small grain size, as well as unconventional resources such as seafloor massive sulfide (SMS) ores, is becoming an important issue [6–8]. Hornn et al. [7] investigated the application of agglomeration using surfactant-stabilized oil emulsion to improve flotation recovery of finely-ground chalcopyrite. The recovery of fine particles ($D_{50} < 5$ μm) by flotation is well known to be difficult due to the low collision probability between air bubbles and mineral particles [9–12]. A study by Hornn et al. [7] revealed that agglomeration using surfactant-stabilized oil emulsion was effective in increasing the apparent particle size of fine chalcopyrite particles from 3.5 μm to ~10 μm, thereby improving Cu recovery from 68% to >97%. Meanwhile, Aikawa et al. [8] proposed a novel flotation procedure to recover chalcopyrite selectively from SMS ores. The major target minerals of SMS ores are chalcopyrite and sphalerite (ZnS), so they should be recovered sequentially via a two-step flotation process whereby chalcopyrite is first recovered, followed by floating sphalerite. In the first stage of flotation, however, both chalcopyrite and sphalerite were recovered together even with zinc sulfate ($ZnSO_4$; sphalerite depressant) because of the presence of anglesite ($PbSO_4$) that releases Pb^{2+} and activates sphalerite. To address this problem, Aikawa et al. [8] employed EDTA washing before flotation, which was effective in removing anglesite, and thus chalcopyrite could be recovered selectively by flotation with $ZnSO_4$.

Apart from studies on the flotation of copper minerals, in this Special Issue, there are two papers on the recovery of fine bauxite [13] and elemental sulfur [14] by flotation. Zhang et al. [13] utilized a plate-packed flotation column (PFC) for fine bauxite, and the Al_2O_3 recovery and grade increased by 2.11% and 1.85%, respectively, compared to the result obtained with unpacked flotation column (UFC). Additionally, they clarified the mechanism of how fine bauxite recovery was improved in a PFC by population balance model (PBM) incorporated with computational fluid dynamics (CFD) techniques; that is, installing the packing-plates significantly reduces the turbulent kinetic energy, which contributed to the formation of small-sized bubbles in the flotation column. A study by Liu et al. [14] demonstrated the recovery of elemental sulfur from the pressure acid leaching residue of ZnS concentrate. The recovery of elemental sulfur from high-sulfur residue can not only add economic value, as it is an important chemical material used for many applications, but also protect the environment. They compared three collectors (e.g., O-Isopropyl-N-Ethyl thionocarbamate (IPETC), ammonium dibutyl dithiophosphate (ADDTP), and sodium ethyl xanthate (SEX)), and among them, IPETC exhibited a superior collecting ability and selectivity to the elemental sulfur, which was further proved by density functional theory (DFT) calculation [14].

2.2. Hydrometallurgy

In this Special Issue, four articles deal with hydrometallurgical processing for Cu, gold (Au), and rare earth metals (REMs) [15–18]. Chae et al. [15] studied hydrochloric acid (HCl) leaching behaviors of Cu and antimony (Sb) in speiss obtained from top submerged lance (TSL) furnace. In general, the speiss containing Cu, Sb, and precious metals (e.g., Au, silver (Ag), etc.) is first processed by sulfuric acid (H_2SO_4) leaching to extract Cu. However, it is reported that the leaching efficiency of Cu from the speiss in H_2SO_4 media is decreased due to the increased amount of Sb in the speiss caused by the co-treatment of various secondary resources during the TSL process. The solubility of Sb in H_2SO_4 is low, which hinders the leaching of Cu. In other words, Sb should be dissolved to improve the leaching efficiency of Cu, so HCl leaching experiments under various conditions were conducted by Chae et al. [15]. As a result, they found the optimum condition of HCl leaching where more than 99% of Cu could be dissolved.

As high-grade ores are depleted, the reprocessing of mine tailings to extract residual valuable metals/minerals is now getting increased attention. Godirilwe et al. [16] studied Cu recovery from mine tailings via flotation, high pressure oxidation leaching (HPOL), and SX-EW process. After flotation of mine tailings, the grade of Cu was upgraded from 0.24% to 0.65%, and the HPOL process of the concentrate yielded a high Cu leaching rate of 94.4%. Afterward, the pregnant leach solution (PLS) containing 2.9 g/L Cu and 102.9 g/L Fe was purified by solvent extraction using LIX-84I, which produced the stripped solution comprising of 44.8 g/L Cu and 1.4 g/L Fe where Cu could be electrowon with a current efficiency of about 95%.

The Cu-catalyzed ammonium thiosulfate leaching is one of the most promising alternatives to cyanidation for extracting gold from ores [17]. A major problem of thiosulfate leaching is its difficulty in recovering gold from the PLS; that is, the conventional Au recovery techniques (e.g., activated carbon (AC) adsorption and cementation using base metals) are inefficient. To address this problem, Jeon et al. [19,20] developed a novel recovery technique called enhanced cementation using AC and zero-valent aluminum (ZVAl), which yielded Au recovery of >99%. However, the previous studies on enhanced cementation using AC and ZVAl were conducted with relatively pure gold thiosulfate solutions [19,20], which significantly differs from the actual PLS where a variety of metal ions coexist. Thus, Jeon et al. [17] investigated the effects of coexisting metal ions (e.g., cobalt (Co), Cu, Fe, nickel (Ni), and zinc (Zn)) on Au recovery via enhanced cementation, and they found that the presence of Fe^{2+}, Co^{2+}, Ni^{2+}, and Zn^{2+} showed the detrimental effects on Au recovery. However, Cu^{2+} acted as a catalyst that minimized the negative effects of these metal ions on Au cementation, yielding 85–90% of Au recovery.

Rare earth metals are essential to achieve a carbon-neutral society because significant quantities of REMs are utilized for manufacturing strong permanent magnets, a critical component used in generators for wind turbines and traction motors for electric vehicles (EVs) [21,22]. There are, however, only a few countries with exploitable rare earth deposits, so the production of REMs through the recycling of wastes is of topical importance. Choubey et al. [18] developed the hydrometallurgical process to extract REMs from Nd-FeB permanent magnets of waste hard disks. The demagnetized magnet was completely dissolved in 2 M H_2SO_4 at 75 °C in 60 min, and then dissolved REMs and Fe were selectively precipitated from the leach liquor at pH 1.75 and 3.5–4.0, respectively. Finally, the precipitated REM hydroxides were converted to their oxides by heating at 120 °C for 2 h.

3. Conclusions and Outlook

A variety of topics have been covered in this Special Issue, presenting recent developments of flotation and leaching. Furthermore, the published articles effectively demonstrate the diversity of the recent research and development in the field. Nevertheless, there are still many challenges to overcome in this research field, so we hope that this Special Issue will serve as a springboard for future discussions and scientific debates on challenging topics related to flotation and leaching.

Having served as a Guest Editor, I am delighted by how well the contributions met the high standards of quality and originality that contributed to the success of this Special Issue. I would like to warmly thank all the authors for their contributions as well as the reviewers for their efforts to ensure a high-quality publication. In addition, I wish to extend my gratitude to the Editors of *Metals* for their continuous support and the *Metals* Editorial Assistants for their valuable and inexhaustible engagement and support during the preparation of this volume. My special thanks go to Mr. Toliver Guo for his support and assistance.

Funding: This research received no external funding.

Conflicts of Interest: The author declares no conflict of interest.

References

1. Hertwich, E.G.; Gibon, T.; Bouman, E.A.; Arvesen, A.; Suh, S.; Heath, G.A.; Bergesen, J.D.; Ramirez, A.; Vega, M.I.; Shi, L. Integrated Life-Cycle Assessment of Electricity-Supply Scenarios Confirms Global Environmental Benefit of Low-Carbon Technologies. *Proc. Natl. Acad. Sci. USA* **2015**, *112*, 6277–6282. [CrossRef]
2. Schlesinger, M.E.; King, M.J.; Sole, K.C.; Davenport, W.G. *Extractive Metallurgy of Copper*, 5th ed.; Elsevier: London, UK, 2011.
3. Park, I.; Hong, S.; Jeon, S.; Ito, M.; Hiroyoshi, N. A Review of Recent Advances in Depression Techniques for Flotation Separation of Cu–Mo Sulfides in Porphyry Copper Deposits. *Metals* **2020**, *10*, 1269. [CrossRef]
4. Park, I.; Hong, S.; Jeon, S.; Ito, M.; Hiroyoshi, N. Flotation Separation of Chalcopyrite and Molybdenite Assisted by Microencapsulation Using Ferrous and Phosphate Ions: Part I. Selective Coating Formation. *Metals* **2020**, *10*, 1667. [CrossRef]
5. Park, I.; Hong, S.; Jeon, S.; Ito, M.; Hiroyoshi, N. Flotation Separation of Chalcopyrite and Molybdenite Assisted by Microencapsulation Using Ferrous and Phosphate Ions: Part II. Flotation. *Metals* **2021**, *11*, 439. [CrossRef]
6. Hornn, V.; Park, I.; Ito, M.; Shimada, H.; Suto, T.; Tabelin, C.B.; Jeon, S.; Hiroyoshi, N. Agglomeration-Flotation of Finely Ground Chalcopyrite Using Surfactant-Stabilized Oil Emulsions: Effects of Co-Existing Minerals and Ions. *Miner. Eng.* **2021**, *171*, 107076. [CrossRef]
7. Hornn, V.; Ito, M.; Shimada, H.; Tabelin, C.B.; Jeon, S.; Park, I.; Hiroyoshi, N. Agglomeration–Flotation of Finely Ground Chalcopyrite Using Emulsified Oil Stabilized by Emulsifiers: Implications for Porphyry Copper Ore Flotation. *Metals* **2020**, *10*, 912. [CrossRef]
8. Aikawa, K.; Ito, M.; Kusano, A.; Park, I.; Oki, T.; Takahashi, T.; Furuya, H.; Hiroyoshi, N. Flotation of Seafloor Massive Sulfide Ores: Combination of Surface Cleaning and Deactivation of Lead-Activated Sphalerite to Improve the Separation Efficiency of Chalcopyrite and Sphalerite. *Metals* **2021**, *11*, 253. [CrossRef]
9. Trahar, W.J. A Rational Interpretation of the Role of Particle Size in Flotation. *Int. J. Miner. Process.* **1981**, *8*, 289–327. [CrossRef]
10. Hornn, V.; Ito, M.; Shimada, H.; Tabelin, C.B.; Jeon, S.; Park, I.; Hiroyoshi, N. Agglomeration-Flotation of Finely Ground Chalcopyrite and Quartz: Effects of Agitation Strength during Agglomeration Using Emulsified Oil on Chalcopyrite. *Minerals* **2020**, *10*, 380. [CrossRef]
11. Hornn, V.; Ito, M.; Yamazawa, R.; Shimada, H.; Tabelin, C.B.; Jeon, S.; Park, I.; Hiroyoshi, N. Kinetic Analysis for Agglomeration-Flotation of Finely Ground Chalcopyrite: Comparison of First Order Kinetic Model and Experimental Results. *Mater. Trans.* **2020**, *61*, 1940–1948. [CrossRef]
12. Bilal, M.; Ito, M.; Koike, K.; Hornn, V.; Ul Hassan, F.; Jeon, S.; Park, I.; Hiroyoshi, N. Effects of Coarse Chalcopyrite on Flotation Behavior of Fine Chalcopyrite. *Miner. Eng.* **2021**, *163*, 106776. [CrossRef]
13. Zhang, P.; Jin, S.; Ou, L.; Zhang, W.; Zhu, Y. Fine Bauxite Recovery Using a Plate-Packed Flotation Column. *Metals* **2020**, *10*, 1184. [CrossRef]
14. Liu, G.; Zhang, B.; Dong, Z.; Zhang, F.; Wang, F.; Jiang, T.; Xu, B. Flotation Performance, Structure-Activity Relationship and Adsorption Mechanism of O-Isopropyl-N-Ethyl Thionocarbamate Collector for Elemental Sulfur in a High-Sulfur Residue. *Metals* **2021**, *11*, 727. [CrossRef]
15. Chae, S.; Yoo, K.; Tabelin, C.B.; Alorro, R.D. Hydrochloric Acid Leaching Behaviors of Copper and Antimony in Speiss Obtained from Top Submerged Lance Furnace. *Metals* **2020**, *10*, 1393. [CrossRef]
16. Godirilwe, L.L.; Haga, K.; Altansukh, B.; Takasaki, Y.; Ishiyama, D.; Trifunovic, V.; Avramovic, L.; Jonovic, R.; Stevanovic, Z.; Shibayama, A. Copper Recovery and Reduction of Environmental Loading from Mine Tailings by High-Pressure Leaching and SX-EW Process. *Metals* **2021**, *11*, 1335. [CrossRef]
17. Jeon, S.; Bright, S.; Park, I.; Tabelin, C.B.; Ito, M.; Hiroyoshi, N. The Effects of Coexisting Copper, Iron, Cobalt, Nickel, and Zinc Ions on Gold Recovery by Enhanced Cementation via Galvanic Interactions between Zero-Valent Aluminum and Activated Carbon in Ammonium Thiosulfate Systems. *Metals* **2021**, *11*, 1352. [CrossRef]
18. Choubey, P.K.; Singh, N.; Panda, R.; Jyothi, R.K.; Yoo, K.; Park, I.; Jha, M.K. Development of Hydrometallurgical Process for Recovery of Rare Earth Metals (Nd, Pr, and Dy) from Nd-Fe-B Magnets. *Metals* **2021**, *11*, 1987. [CrossRef]
19. Jeon, S.; Tabelin, C.B.; Takahashi, H.; Park, I.; Ito, M.; Hiroyoshi, N. Enhanced Cementation of Gold via Galvanic Interactions Using Activated Carbon and Zero-Valent Aluminum: A Novel Approach to Recover Gold Ions from Ammonium Thiosulfate Medium. *Hydrometallurgy* **2020**, *191*, 105165. [CrossRef]
20. Jeon, S.; Bright, S.; Park, I.; Tabelin, C.B.; Ito, M.; Hiroyoshi, N. A Simple and Efficient Recovery Technique for Gold Ions from Ammonium Thiosulfate Medium by Galvanic Interactions of Zero-Valent Aluminum and Activated Carbon: A Parametric and Mechanistic Study of Cementation. *Hydrometallurgy* **2021**, *208*, 105815. [CrossRef]
21. Park, I.; Kanazawa, Y.; Sato, N.; Galtchandmani, P.; Jha, M.K.; Tabelin, C.B.; Jeon, S.; Ito, M.; Hiroyoshi, N. Beneficiation of Low-Grade Rare Earth Ore from Khalzan Buregtei Deposit (Mongolia) by Magnetic Separation. *Minerals* **2021**, *11*, 1432. [CrossRef]
22. Jha, M.K.; Choubey, P.K.; Dinkar, O.S.; Panda, R.; Jyothi, R.K.; Yoo, K.; Park, I. Recovery of Rare Earth Metals (REMs) from Nickel Metal Hydride Batteries of Electric Vehicles. *Minerals* **2022**, *12*, 34. [CrossRef]

 *metals*

Review

A Review of Recent Advances in Depression Techniques for Flotation Separation of Cu–Mo Sulfides in Porphyry Copper Deposits

Ilhwan Park [1,*], Seunggwan Hong [2], Sanghee Jeon [1], Mayumi Ito [1] and Naoki Hiroyoshi [1]

[1] Division of Sustainable Resources Engineering, Faculty of Engineering, Hokkaido University, Sapporo 060-8628, Japan; shjun1121@eng.hokudai.ac.jp (S.J.); itomayu@eng.hokudai.ac.jp (M.I.); hiroyosi@eng.hokudai.ac.jp (N.H.)
[2] Division of Sustainable Resources Engineering, Graduate School of Engineering, Hokkaido University, Sapporo 060-8628, Japan; chanitroi@gmail.com
* Correspondence: i-park@eng.hokudai.ac.jp; Tel.: +81-11-706-6315

Received: 29 August 2020; Accepted: 16 September 2020; Published: 21 September 2020

Abstract: Porphyry copper deposits (PCDs) are some of the most important sources of copper (Cu) and molybdenum (Mo). Typically, the separation and recovery of chalcopyrite ($CuFeS_2$) and molybdenite (MoS_2), the major Cu and Mo minerals, respectively, in PCDs are achieved by two-step flotation involving (1) bulk flotation to separate Cu–Mo concentrates and tailings (e.g., pyrite, silicate, and aluminosilicate minerals) and (2) Cu–Mo flotation to separate chalcopyrite and molybdenite. In Cu–Mo flotation, chalcopyrite is depressed using Cu depressants, such as NaHS, Na_2S, Nokes reagent (P_2S_5 + NaOH), and NaCN, meaning that it is recovered as tailings, while molybdenite is floated and recovered as froth product. Although conventionally used depressants are effective in the separation of Cu and Mo, they have the potential to emit toxic and deadly gases such as H_2S and HCN when operating conditions are not properly controlled. To address these problems caused by the use of conventional depressants, many studies aimed to develop alternative methods of depressing either chalcopyrite or molybdenite. In this review, recent advances in chalcopyrite and molybdenite depressions for Cu–Mo flotation separation are reviewed, including alternative organic and inorganic depressants for Cu or Mo, as well as oxidation-treatment technologies, such as ozone (O_3), plasma, hydrogen peroxide (H_2O_2), and electrolysis, which create hydrophilic coatings on the mineral surface.

Keywords: porphyry copper deposits; chalcopyrite; molybdenite; flotation; conventional and alternative depression techniques

---

1. Introduction

Porphyry copper deposits (PCDs) are the world's most important sources of copper (Cu) because they account for more than 60% of global annual copper production [1,2]. Although these deposits are widespread, they seem to be localized in time and space within the overall evolutionary pattern of magmatic arcs along plate convergent margins; that is, they are predominantly associated with (i) Mesozoic to Cenozoic orogenic belts in western North and South America, (ii) Tethyan orogenic belts in eastern Europe and southern Asia, and (iii) Paleozoic orogens in Central Asia and eastern North America (Figure 1) [1,3–5]. PCDs are low-grade (around 0.3–2.0% Cu; average 0.44% copper in 2008); however, they have significant economic value due to their large size (typically greater than 100 million tons), long mine lives (spanning several decades), and high production rates (i.e., millions of tons of copper per year) [1]. These deposits are also important sources of molybdenum (Mo), gold (Au), and silver (Ag) [1,5]. PCDs greatly contribute to the world's supply of Mo, accounting for more than 50% of the total supply [6]. In addition, elevated concentrations of rhenium (Re), tellurium (Te), and platinum-group elements (PGEs) are incorporated in PCDs, which are recovered as byproducts; that is, Re, Te, and PGEs are mostly recovered from processes related to molybdenite (MoS_2), anode slimes generated from the electrorefining of copper anodes, and the smelting of copper ores, respectively [1].

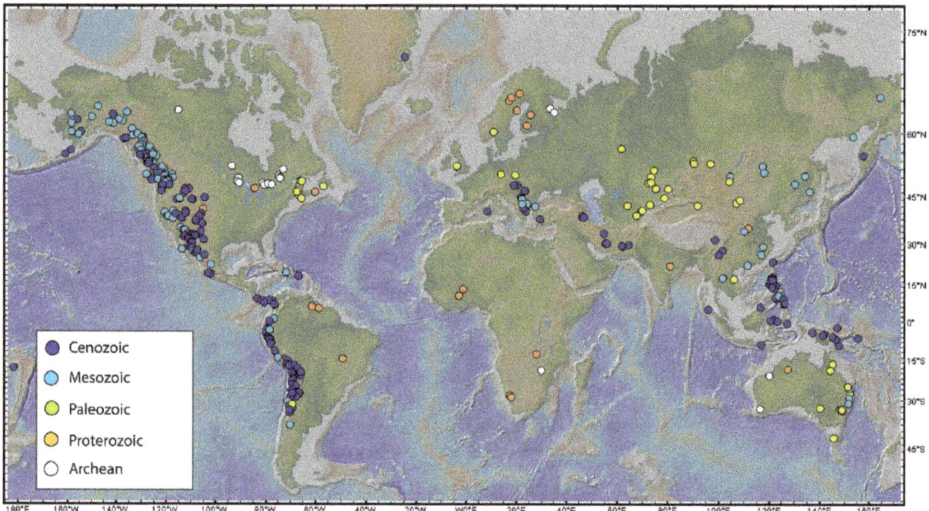

Figure 1. Occurrence of porphyry copper deposits through time (reprinted with permission from Lee and Tang [3], copyright (2020) Elsevier).

Typically, the recovery of copper sulfides (mostly chalcopyrite) and molybdenite from PCDs is conducted via a series of processes; that is, open-pit mining of PCDs to excavate the ores, closed-circuit comminution to liberate valuable and nonvaluable minerals, and multistep flotation of ground ores to separate Cu–Mo concentrates and tailings (e.g., pyrite and silicate minerals), followed by the separation of Cu and Mo minerals from Cu–Mo concentrates by flotation using Cu depressants (e.g., NaHS, Na_2S, Nokes reagent (P_2S_5 + NaOH), and NaCN)—the detailed processes for PCDs are discussed in the following section. Afterwards, Cu concentrates are treated by a pyrometallurgical process to produce Cu metal, while Mo concentrates are treated by leaching using HCl–$FeCl_3$ to produce the high-grade MoS_2 used for lubricants, or by roasting and acid pressure-oxidation processes to produce technical Mo oxide (MoO_3) [7].

During Cu–Mo flotation separation, Cu depressants work effectively to separate Cu and Mo minerals, but there are serious drawbacks, such as (i) the potential to release toxic and deadly gases (e.g., H_2S and HCN) when pulp pH is not properly maintained; (ii) the corrosive nature of Cu depressants, which destroy pipelines; (iii) the imperfect recovery of molybdenite; and (iv) considerable losses of precious metals such as gold and silver when cyanide is used [8–10]. In order to replace conventionally used Cu depressants, which have the above limitations, many studies have been conducted to develop alternative methods, including the use of environmentally friendly organic and inorganic depressants to reduce the floatability of either chalcopyrite or molybdenite, and oxidation treatments involving the use of ozone (O_3), plasma, hydrogen peroxide (H_2O_2), and electrolysis to create hydrophilic coatings on the surfaces of the chalcopyrite. Despite the existence of extensive studies, there is no review that summarizes all findings on this important topic. In this review, therefore, newly developed techniques for Cu–Mo flotation separation are reviewed. Specifically, we discuss mechanisms involved in selective depression for either chalcopyrite or molybdenite, and the advantages and disadvantages of each technique.

2. Typical Process of Cu–Mo Sulfide Ores

Figure 2 shows the typical flow sheet for the beneficiation of Cu–Mo sulfide ores, consisting of three steps: (1) comminution to liberate target minerals, (2) bulk flotation to recover Cu–Mo minerals from gangue minerals, and (3) selective flotation of Mo minerals (mostly molybdenite) from Cu–Mo bulk concentrates [10]. The ores are first crushed by a primary crusher, such as a jaw or gyratory crusher, and then ground with a semi-autogenous (SAG) and/or ball mill where conditioning agents such as a Cu–Mo collector (e.g., xanthate- and/or oil-based collector(s)) and pH adjuster (e.g., lime (CaO)) are introduced to improve the separation of Cu–Mo and gangue minerals during bulk flotation. The purpose of adding lime is to make the pulp pH alkaline, at which pyrite, one of the representative gangue minerals, could be effectively depressed. This depression of pyrite under alkaline conditions could be explained by the competitive adsorption of OH^- and the xanthate-based collector. There are critical pH values for sulfide minerals, below which xanthate ion can be adsorbed, while above which its adsorption is inhibited due to the competitive adsorption of OH^- [11]. For instance, critical pH values for pyrite and chalcopyrite in the solution containing 25 mg/L potassium ethyl xanthate at room temperature are 10.5 and 11.8, respectively, so the selective flotation of chalcopyrite could be achieved when pulp pH is adjusted to be between 10.5 and 11.8 [11]. There are two additional mechanisms of pyrite depression under alkaline conditions: (i) dixanthogen, an adsorbed form of xanthate-based collector on pyrite surface, becomes thermodynamically unstable; and (ii) pyrite surface is covered with ferric hydroxide having a hydrophilic nature [12]. After bulk flotation, tailings are disposed of into the tailings storage facility (TSF), while concentrates are transferred to the conditioning process, followed by flotation for the separation of Cu and Mo minerals. As shown in Figure 2, the conditioning process is aimed at depressing Cu minerals by using Cu depressants such as NaHS. This depressant is readily dissociated in the aqueous solution, and produces NaOH and H_2S (Equation (1)). At pH 9–10, H_2S is transformed into HS^- (Equation (2)), which reacts with xanthate-adsorbed Cu minerals; thereby, the surfaces of Cu minerals are modified from hydrophobic to hydrophilic due to the desorption of xanthate adsorbed on them (Equation (3), where CuX denotes the surface of xanthate-adsorbed Cu minerals). After pretreating Cu–Mo bulk concentrates with NaHS, Mo minerals are recovered as froth products, while Cu minerals are recovered as tailings. According to Hirajima et al. [13], the pretreatment of Cu–Mo bulk concentrates using NaHS significantly decreased Cu recovery from 85 to 10%, while Mo recovery was increased from 85 to 99% (Figure 3), indicating that NaHS was effective in selectively depressing Cu minerals. As a result of a single stage of Cu–Mo flotation using NaHS, the froth product contains around 10% Cu, which is most likely caused by entrainment. In industry, thus, froth products are further processed via multiple cleaning stages to meet the requirement of saleable Mo concentrates.

$$NaHS + H_2O \leftrightarrow NaOH + H_2S \tag{1}$$

$$H_2S \leftrightarrow H^+ + HS^- \tag{2}$$

$$2CuX + HS^- \leftrightarrow Cu_2S + 2X^- + H^+ \tag{3}$$

Similarly, the Nokes reagent (P_2S_5 + NaOH), Na_2S, and NaCN were also adopted as Cu depressants [7,10,14,15]. The function of the Nokes reagent and Na_2S is the same as that of NaHS for producing HS^-, which desorbs the adsorbed xanthate on Cu minerals (Equations (2)–(6)).

$$P_2S_5 + 6NaOH \leftrightarrow 2Na_3PO_2S_2 + H_2S + 2H_2O \tag{4}$$

$$P_2S_5 + 10NaOH \leftrightarrow Na_3PO_2S_2 + Na_3PO_3S + 2Na_2S + 5H_2O \tag{5}$$

$$Na_2S + 2H_2O \leftrightarrow H_2S + 2NaOH \tag{6}$$

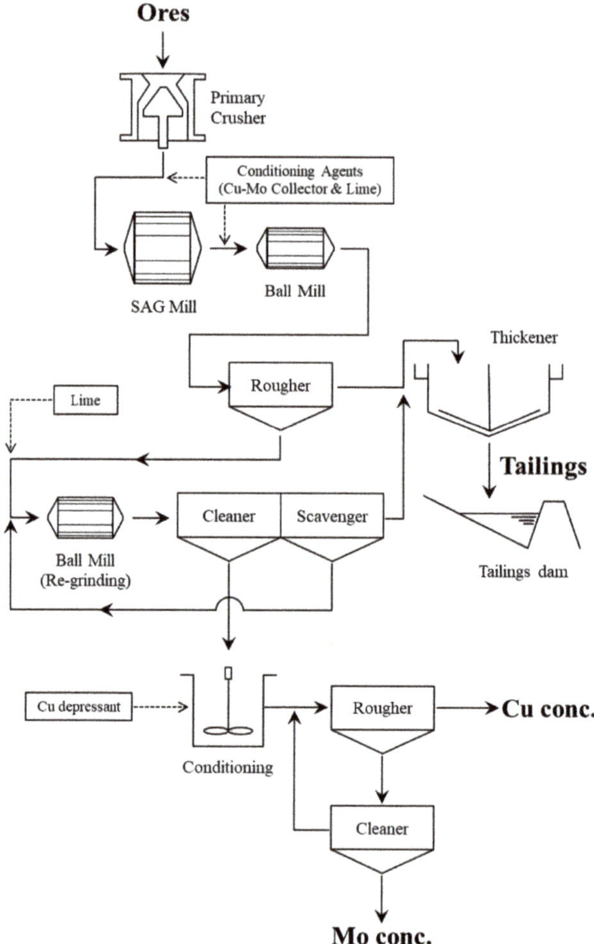

Figure 2. Beneficiation flowsheet of Cu–Mo sulfide ores [7].

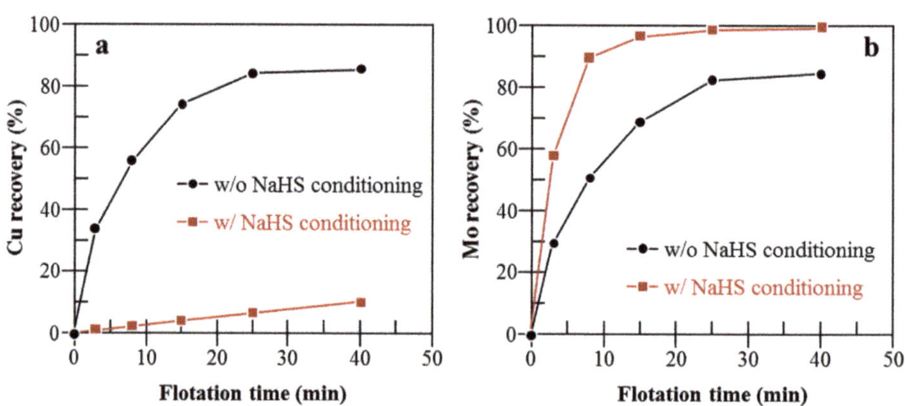

Figure 3. Effects of NaHS on recovery of (**a**) chalcopyrite and (**b**) molybdenite (reprinted with permission from Hirajima et al. [13], copyright (2017) Elsevier).

In the case of NaCN, its depressive mechanism is different, as NaCN is hydrolyzed in aqueous solution to form NaOH and HCN (Equation (7)). Then, the latter dissociates to CN^- under alkaline conditions (Equation (8)). The action of CN^- as a Cu depressant lies in its strong ability to form a copper–cyanide complex (Equation (9)) [14].

$$NaCN + H_2O \leftrightarrow NaOH + HCN \tag{7}$$

$$HCN \leftrightarrow H^+ + CN^- \tag{8}$$

$$2Cu^{2+} + 6CN^- \leftrightarrow 2[Cu(CN)_2]^- + C_2N_2 \tag{9}$$

By increasing the concentration of CN^-, it creates additional species of the copper–cyanide complex such as $[Cu(CN)_3]^{2-}$ and $[Cu(CN)_4]^{3-}$ [16]. Cyanide reacts with not only Cu^{2+}, but also copper–xanthate formed on the surfaces of Cu minerals, resulting in a decrease in their hydrophobicity due to the replacement of xanthate with cyanide. In addition, cyanide can directly react with the surface of Cu minerals where CN^- is adsorbed, thereby making it impossible to adsorb xanthate [16], and it is a reducing agent that reduces pulp potential in which chalcopyrite does not float.

Although effective, these depressants typically used for Cu–Mo separation have the potential to generate toxic and lethal gases if used haphazardly. As shown in Figure 4, the protonated forms of H_2S and HCN start forming at a pH below 10 and 12, respectively. Once they exist in aqueous solution, their vaporizations are readily progressed even under ambient conditions due to their relatively high vapor pressure (i.e., P_{H2S} = 20.03 atm at 25 °C; P_{HCN} = 0.98 atm at 25 °C) [17,18]. The problem of the formation of vaporized H_2S and HCN lies in its serious toxicity to human beings, the toxic actions of which occur via the inhibition of cytochrome oxidase that prevents the cellular utilization of oxygen, followed by the inhibition of the terminal step of electron transport in brain cells, resulting in loss of consciousness, respiratory arrest, and ultimately death [19,20]. Another problem of using NaCN in flotation is the significant losses of precious metals such as gold and silver incorporated in PCDs because cyanide is known to dissolve them by forming stable complexes [10]. Because of these problems, mineral-processing plants in which conventional depressants are used should either consist of covered flotation cells with an active ventilation system or always maintain pulp pH at above around 9.5 [21].

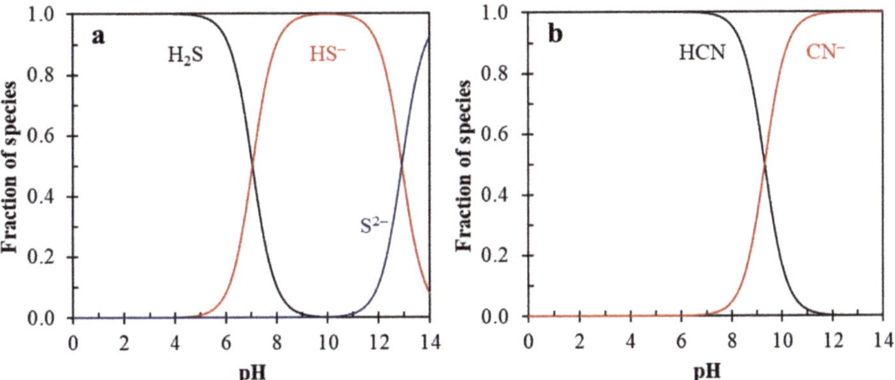

Figure 4. pH-dependent speciation of (**a**) sulfide (pK_{a1} = 7.1 and pK_{a2} = 12.9; $[S]_{tot}$ = 1 M) and (**b**) cyanide (pK_a = 9.3; $[CN]_{tot}$ = 1 M).

3. Alternative Options for Selective Flotation of Cu–Mo Bulk Concentrates

3.1. Molybdenite Depression

To replace the use of potentially toxic depressants such as NaHS and NaCN, there have been significant efforts to find suitable alternative depressants for molybdenite, for example, dextrin, lignosulfonate, carboxymethyl-based organic compounds, and humic acid. The summary of this section is present in Table 1.

Table 1. Summary of molybdenite depression.

Mo Depressant	Feed	Results	Features
Dextrin [22–26]	Natural MoS_2 [24], Cu ore (0.7–3.03% Cu, 0.011–0.14% Mo) [26].	✓ Mo recovery (R_{Mo}) decreased from 92% to 1.5% with 100 mg/L dextrin [24]. ✓ Recovery of Cu (R_{Cu}) and Mo (R_{Mo}) was 88.2% and 18.5%, respectively, with 200 g/t dextrin [26].	✓ In the presence of organic compound (e.g., iso-octane), the depressing efficiency of dextrin decreased [24]. ✓ Dextrin is effective in depressing MoS_2, but increases water recovery, causing the recovery of unwanted minerals [26].
Lignosulfonate [27–31]	Natural $CuFeS_2$ and MoS_2 [28].	✓ Single mineral-flotation tests showed that sodium-based lignosulfonates were effective in selectively depressing MoS_2 [28].	✓ Presence of Ca^{2+} introduced by calcium-based lignosulfonates and/or lime (pH adjuster) to the flotation system depressed not only MoS_2 but also $CuFeS_2$ [28].
O-carboxymethyl chitosan (O-CMC) [32,33]	Natural $CuFeS_2$ and MoS_2 [32].	✓ Selective depression of MoS_2 (R_{Mo} < 12%; R_{Cu} > 90%) was achieved using 150 ppm O-CMC with 20 ppm potassium isobutyl xanthate (KIBX) and 20 ppm methyl isobutyl carbinol (MIBC) at pH 5–9 [32].	✓ O-CMC could be adsorbed on both minerals but via different mechanisms; O-CMC was adsorbed on $CuFeS_2$ via weak physical interactions, while its adsorption on MoS_2 occurred via hydrophobic interactions [32,33].
Humic acid (HA) [34]	Natural $CuFeS_2$ and MoS_2 [34].	✓ Mixed-mineral flotation with 20 ppm HA resulted in >80% R_{Cu} and $R_{Mo} \approx 20$% at pH 5–9 [34].	✓ Similar to O-CMC, the adsorption mechanisms of HA/$CuFeS_2$ and HA/MoS_2 were defined as electrostatic and hydrophobic interactions, respectively [34].
Carboxymethylcellulose (CMC) [35–37]	Natural MoS_2 [35].	✓ Two CMC polymers (i.e., HSHB and LSLB) were tested, and both were effective in depressing MoS_2; in SPW, R_{Mo} was 8% with 5 ppm HSHB and 2% with 5 ppm LSLB [35].	✓ CMCs act as both a Mo and a Cu depressant [36,37]. ✓ Depending on the type of ore, two approaches (i.e., (1) depressing MoS_2 with floating $CuFeS_2$ and (2) depressing $CuFeS_2$ with floating MoS_2) can be applicable [35–37].

3.1.1. Dextrin

Hernlund [22], for example, used dextrin as a Mo depressant. Dextrin is a water-soluble polysaccharide having the general formula of $(C_6H_{10}O_5)_n$, produced by the enzymatic hydrolysis of starch [23,24]. The adsorption mechanism of dextrin molecules on the surface of molybdenite was proposed to occur via hydrophobic interaction [24,25]. As a result, the molybdenite surface is rendered hydrophilic, so its recovery decreased significantly from 92% to 1.5% when 100 mg/L dextrin was added [24]. A similar result was obtained by Jorjani et al. [26], who investigated the flotation behavior of porphyry copper ores containing Cu minerals (e.g., chalcopyrite and chalcocite), molybdenite, and aluminosilicates (e.g., albite, illite, kaolinite, muscovite, orthoclase, and vermiculite). This depressant could work well to decrease molybdenite recovery from around 50% to 15%; however, at a certain amount of added dextrin (i.e., 200 g/ton), it causes an increase in water content in froth products, resulting in the recovery of unwanted gangue minerals such as aluminosilicates by entrainment. In other words, it is recommended for froth products to be washed to obtain better products with low contents of undesired minerals [26]. Moreover, dextrin could not act as a Mo depressant in the presence of oily collectors, e.g., iso-octane (2,2,4-trimethylpentane, $(CH_3)_3CCH_2CH(CH_3)_2$), which is another disadvantage of the use of dextrin for Cu–Mo flotation [24].

3.1.2. Lignosulfonates

Lignosulfonates, a group of water-soluble and strong anionic polyelectrolytes typically obtained as a byproduct of wood processing for the extraction of cellulose [27,28], were first used for the separation of molybdenite and talc, both of which are hydrophobic in nature [29–31]. For molybdenite/talc separation, the reverse flotation of talc was proposed, where talc-bearing molybdenite ores are conditioned by lignosulfonate, with lime used as pH adjuster to increase the pH to around 11.5, resulting in the selective depression of molybdenite while talc is floated. In the case of chalcopyrite/molybdenite separation, however, the combination of lignosulfonate and lime depressed not only molybdenite but also chalcopyrite at pH ~11, making their separation impossible [28]. The floatability of chalcopyrite at the same pH but adjusted using KOH was unaffected by the presence of lignosulfonates. From these results, it can be concluded that Ca^{2+} has a strong effect on chalcopyrite depression, especially under alkaline conditions most likely due to the formation of $Ca(OH)_2$ on its surface. There are two possible mechanisms of how the formation of $Ca(OH)_2$ depresses the floatability of chalcopyrite: (1) $Ca(OH)_2$ is a hydrophilic compound, so it directly prevents the bubble attachment; and (2) the formation of $Ca(OH)_2$ changes the surface charge of chalcopyrite from negative to positive, making lignosulfonates (i.e., strong anionic polyelectrolytes) favorable to be adsorbed [27,28]. On the other hand, the floatability of molybdenite was strongly depressed by lignosulfonate in a wide pH range of 5–11, regardless of used pH adjusters (e.g., CaO, Na_2CO_3, and KOH). These results suggest that chalcopyrite/molybdenite separation could be achieved by using (not calcium-based) sodium sulfonates in the absence of Ca^{2+}. This means that lime, the most commonly used pH adjuster for depressing pyrite in bulk Cu/Mo flotation, is required to be replaced with other basic materials to eliminate the presence of Ca^{2+} in the pulp. Similar to the case of dextrin, lignosulfonates also lose their ability to depress molybdenite when the mineral surface is rendered hydrophobic by an oily collector such as dodecane ($CH_3(CH_2)_{10}CH_3$) prior to depressant adsorption [28].

3.1.3. Carboxymethyl-Based Organic Compounds and Humic Acid

O-carboxymethyl chitosan (O-CMC), a derivative of the second most abundant natural polysaccharide (i.e., chitosan), is nontoxic, biodegradable, cost-effective, and has better solubility in water compared to that of chitosan, which make it suitable for uses in a wide range of technologies. Yuan and coworkers [32,33] utilized O-CMC for the depression of molybdenite during Cu–Mo flotation. As shown in Figure 5a, the result of single-mineral flotation with 20 ppm potassium isobutyl xanthate (KIBX) showed that both minerals were highly recoverable (~97%) in the absence of O-CMC; however, with the addition of 150 ppm O-CMC, molybdenite was significantly depressed, and its recovery decreased to around 11%, whereas the floatability of chalcopyrite was not affected by O-CMC [32]. This selective depression of molybdenite was also achieved during the flotation of a Cu–Mo mixture in the presence of 150 ppm O-CMC (Figure 5b). The topographic atomic-force-microscopy (AFM) images of chalcopyrite and molybdenite before and after treatment with O-CMC are illustrated in Figure 5c–f, showing that O-CMC created aggregates (100–200 nm diameter with ~2 nm height) that were only present on the surface of molybdenite. This result implies that O-CMC cannot be adsorbed on the chalcopyrite surface, but it is indeed possible according to their follow-up study [33]. Electrokinetic studies showed that the zeta potential of chalcopyrite treated with O-CMC notably decreased and was close to that of O-CMC macromolecules. In addition, X-ray photoelectron-spectroscopy (XPS) measurements of O-CMC-treated chalcopyrite indicated that apparent signatures of O-CMC were detected in the spectra of C 1s, N 1s, and O 1s. Although O-CMC could be adsorbed on the chalcopyrite surface, its adsorption is reversible due to the weak physical interactions (e.g., electrostatic interactions), which means that it could be mechanically desorbed. After washing O-CMC-treated chalcopyrite with Milli-Q water, in fact, the signals of adsorbed O-CMC disappeared in AFM and time-of-flight secondary ion-mass spectrometry (ToF-SIMS) imaging and diffused-reflectance infrared Fourier transform (DRIFT) spectroscopy [33]. However, O-CMC is adsorbed on the molybdenite basal planes via hydrophobic interactions that are more irreversible than those between O-CMC and chalcopyrite. Due to the different adsorption characteristics of O-CMC on chalcopyrite/molybdenite, selective depression of molybdenite could be achieved during Cu–Mo flotation. Yuan and coworkers [34] investigated the selective depression of humic acid (HA), a major organic constituent of soil and one of the most abundant naturally occurring organic macromolecules, for Cu–Mo flotation, and their results indicated that HA could selectively depress molybdenite by a similar adsorption mechanism as that of O-CMC; that is, the adsorption of HA on chalcopyrite and molybdenite takes place via electrostatic and hydrophobic interactions, respectively.

In the case of polymer adsorption on the mineral surface, electrolyte concentration and composition had considerable impact on adsorbed layer properties (e.g., thickness, coverage, and roughness) [35]. For example, Kor et al. [35] investigated the effects of electrolyte concentration and composition on the adsorption of two carboxymethylcellulose (CMC) polymers (high substitution, high blockiness (HSHB); low substitution, low blockiness (LSLB)) onto the molybdenite surface, and confirmed that higher ionic strength (2.76×10^{-2} M KCl) contributes to thicker layers with higher coverage compared to those observed in 10^{-2} M KCl (Table 2). Moreover, simulated process water (SPW), a complex electrolyte containing multivalent metal ions, further increases the thickness of adsorbed layers. As shown in Figure 6a, the flotation recovery of bare molybdenite in 10^{-2} M KCl electrolyte is around 92%, but the addition of 5 mg/L HSHB decreased Mo recovery to 30% in 10^{-2} M KCl, 14% in 2.76×10^{-2} M KCl, and 8% in SPW. Compared with HSHB, the depressing effect of LSLB for molybdenite is stronger, that is, the recovery of LSLB-treated molybdenite was 5% in in 10^{-2} M KCl, 4% in 2.76×10^{-2} M KCl, and 2% in SPW (Figure 6b), indicating that a thicker layer with higher coverage results in the better suppression of molybdenite floatability.

However, the floatability of chalcopyrite is also affected by HSHB and LSLB. In the absence of CMCs, the maximal recovery (R_{max}) of chalcopyrite is 91% ± 5%, whereas R_{max} is decreased to 66% ± 4% with 25 mg/L HSHB and 38% ± 2% with 25 mg/L LSLB [36]. According to Qui et al. [37], CMCs can be used for depressing chalcopyrite in Cu–Mo flotation separation. This means that the depressing ability of CMCs is not limited to molybdenite, that is, CMCs can play roles in depressing chalcopyrite and/or molybdenite, which is strongly dependent on operating conditions and ore compositions. Similarly, the problems with the use of organic polymers (e.g., starches and dextrins), widely used as a Pb depressant for Cu–Pb separation and as a pyrite depressant, are associated with the nonspecific depression of all sulfides when an excessive dosage is introduced [38]; thus, the flotation circuits using organic depressants need constant attention to avoid failure in sulfide separation.

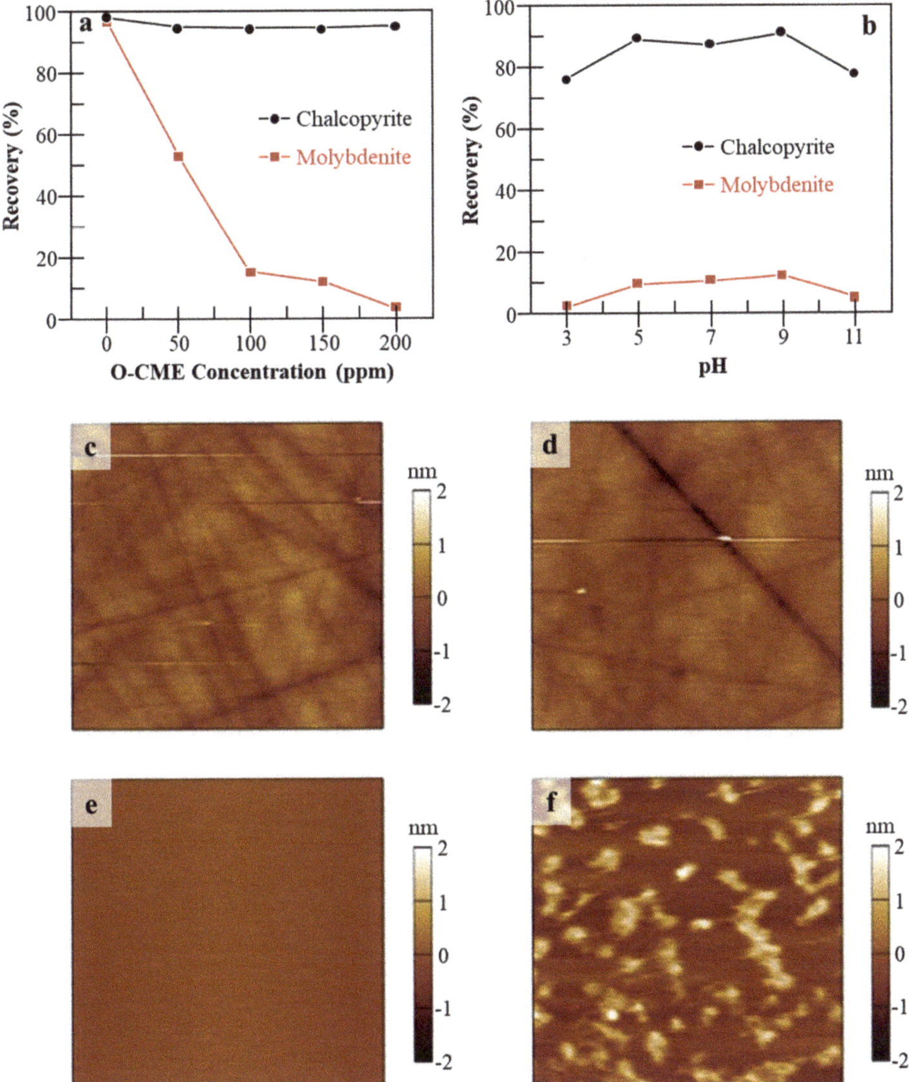

Figure 5. Effect of O-CMC on selective depression of molybdenite: (**a**) single-mineral flotation in 1 mM KCl solution with 20 ppm KIBX at pH 9 under various concentrations of O-CMC; (**b**) flotation of artificially mixed minerals (chalycopyrite:molybdenite = 1:1 by weight) in 1 mM KCl solution with 20 ppm KIBX and150 ppm O-CMC as function of pH, and atomic-force-microscopy (AFM) height profiles of (**c**) bare chalcopyrite, (**d**) chalcopyrite treated with 150 ppm O-CMC, (**e**) bare molybdenite, and (**f**) molybdenite treated with 150 ppm O-CMC (reprinted with permission from Yuan et al. [32], copyright (2019) Elsevier).

Table 2. Thickness and surface coverage (Γ) of adsorbed CMC (5 mg/L) at pH 9 on molybdenite (reprinted with permission from Kor et al. [35], copyright (2014) American Chemical Society). HSHB, high substitution, high blockiness; LSLB, low substitution, low blockiness; SPW, simulated process water.

Electrolyte	HSHB		LSLB	
	Thickness (nm)	Γ (%)	Thickness (nm)	Γ (%)
10^{-2} M KCl	1.1 ± 0.2	23	2.6 ± 0.2	100
2.76×10^{-2} M KCl	1.6 ± 0.6	95	4.0 ± 0.4	100
SPW *	2.7 ± 0.5	100	6.3 ± 0.4	100

* SPW consisted of 2.745 mM $CaSO_4$, 0.411 mM $Mg(NO_3)_2$, 4.653 mM Na_2SO_4, and 1.458 mM KCl.

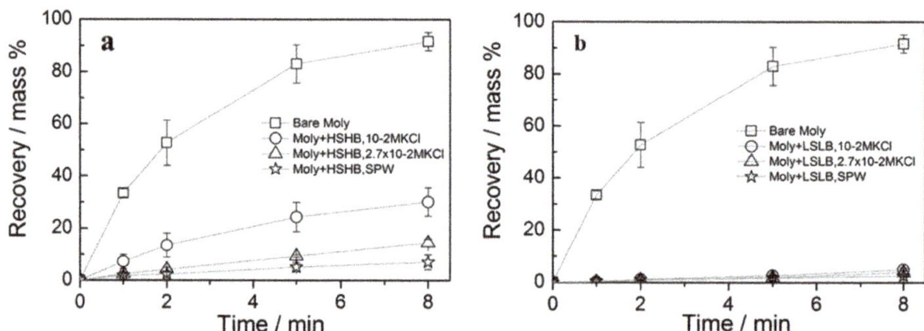

Figure 6. Single-mineral flotation of bare molybdenite and in presence of 5 mg/L CMC ((**a**) HSHB and (**b**) LSLB) at three studied electrolyte conditions: 10^{-2} M KCl, 2. 76 × 10^{-2} M KCl, and SPW, all at pH 9 (reprinted with permission from Kor et al. [35], copyright (2014) American Chemical Society).

3.2. Depression of Cu Minerals

As discussed in an earlier section, many molybdenite depressants could effectively improve the separation between Mo and Cu minerals. In the case of PCDs, however, the strategy of depressing chalcopyrite has been more commonly adopted than depressing molybdenite has, primarily due to mass-balance considerations [28], that is, the amount of molybdenite in PCDs is typically lower than that of Cu minerals, which makes the flotation process that recovers molybdenite by depressing chalcopyrite attractive. If the separation is done by the depression of molybdenite and the simultaneous flotation of chalcopyrite, it may cause the mechanical entrainment of molybdenite within a large volume of chalcopyrite concentrate [28], which lowers the grade of Cu concentrate, and leads to appreciable loss of molybdenite. Alternative depression techniques for chalcopyrite, including inorganic/organic depressants and oxidation treatments, are reviewed and summarized in Table 3.

Table 3. Summary of chalcopyrite depression.

Cu Depressant	Feed		Results		Features
Na_2SO_3 [39–45]	Natural $CuFeS_2$ and MoS_2 [44].	✓	With the addition of 0.1 M Na_2SO_3, $CuFeS_2$ floatability dramatically decreased to ≈ 0%, while >95% MoS_2 floated [44].	✓	Na_2SO_3 can also act as an activator for $CuFeS_2$, of which the surface is covered with ferric oxyhydroxide [45].
Seawater [46–58]	Natural $CuFeS_2$ and MoS_2 [49].	✓	Mixed-mineral flotation using artificial seawater at pH 10 resulted in lower recovery of both minerals (around 10–15%), but separation efficiency was greatly improved by using 416 mg/L emulsified kerosene (R_{Cu} > 70%, R_{Mo} ≈ 20%) [49].	✓	The adsorption of seawater precipitates (e.g., $Mg(OH)_2$ and $CaCO_3$) is the primary cause of both minerals' depression; however, emulsified oil limits its adsorption on MoS_2 [49].
Organic depressants (e.g., PGA [59], DMSA [60], DBT [61], chitosan [62–65], ATDT [66], and AHS [67])	Bulk Cu–Mo conc. [59], Cu–Mo ore [60], Cu–Mo rough conc. [61], natural $CuFeS_2$ and MoS_2 [62], $CuFeS_2$ and MoS_2 purified by flotation [66,67].	✓	All reviewed organic depressants were effective in depressing $CuFeS_2$, while they were negligible for MoS_2 floatability [59–62,66,67].	✓	S and/or N atoms in organic compounds have strong affinity with Cu atoms in $CuFeS_2$, making these organic depressants highly selective for $CuFeS_2$ [59–62,66,67].
Oxidation treatments (e.g., ozone (O_3) [68–73], plasma [74], H_2O_2 [13,75–77], and electrolysis [78]).	Bulk Cu–Mo conc. [69], natural $CuFeS_2$ and MoS_2 [13,74,75], mineral electrodes [78].	✓	After oxidation treatments, chalcopyrite is aggressively oxidized, resulting in a decrease in $CuFeS_2$ floatability due to the formation of hydrophilic oxidation products on its surface [13,69,74,75,78].	✓	MoS_2 was also oxidized, and on its surface, oxidation products (e.g., MoO_3) were formed; however, MoO_3 is highly soluble under alkaline conditions, so its effect on MoS_2 floatability becomes negligible under typical Cu–Mo flotation conditions (pH > 9) [13,74,75,78].

3.2.1. Inorganic Depressants

The xanthate-induced flotation of sulfide minerals is significantly influenced by pulp potential [39–42]. For example, the flotation of chalcopyrite using butyl xanthate shows that its recovery was lowered with decreasing Eh (Figure 7). At the Eh of −0.4 V, the addition of a collector had a negligible effect on the recovery of chalcopyrite, but was effective in the Eh range from −0.1 to 0.2 V due to the formation of a Cu(I)–xanthate complex (−0.1 to 0.0 V; Equation (10)) as well as dixanthogen (0.0 to 0.1 V; Equation (11)) [42].

$$CuX + FeS_2 + e^- \rightarrow CuFeS_2 + X^-, E^0 = -0.096 \text{ V} \tag{10}$$

$$X_2 + 2e^- \rightarrow 2X^-, E^0 = -0.009 \text{ V} \tag{11}$$

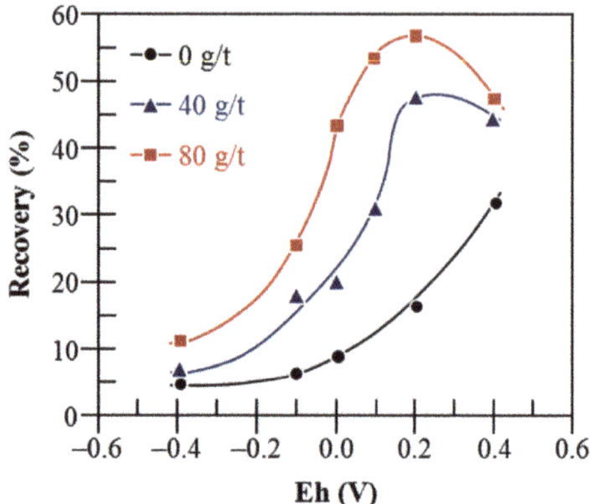

Figure 7. Recovery of chalcopyrite as function of Eh and butyl xanthate addition at pH 8.1 (reprinted with permission from Grano et al. [39], copyright (1990) Elsevier).

At an Eh greater than 0.2 V, however, the recovery of chalcopyrite started decreasing because of the decomposition of the Cu(I)–xanthate complex (Equation (12)), making the mineral surface less hydrophobic [43].

$$2HCuO_2^- + X_2 + 6H^+ + 4e^- \rightarrow 2CuX + 4H_2O, E^0 = -1.402 \text{ V} \tag{12}$$

On the basis of Eh dependence on chalcopyrite flotation, Miki et al. [44] used Na_2SO_3 as a depressant that provided the reducing conditions where the floatability of chalcopyrite decreased. The flotation results using a TMD solution (mixture of TX15216 (alkyl mercaptan), MX-7017 (modified thionocarbamate), and diesel oil), and methyl isobutyl carbinol (MIBC) as collector and frother, respectively, indicated that molybdenite was recovered by more than 95%, while the recovery of chalcopyrite was almost 0% when conditioned with 0.1 M Na_2SO_3 at pH 10.8 for 1 h (Figure 8a). XPS spectra of chalcopyrite treated with Na_2SO_3 showed that various hydrophilic species (i.e., CuO, $Cu(OH)_2$, FeOOH, and $Fe_2(SO_4)_3$) were formed on its surface. Miki and coworkers [44] proposed the mechanism on the formation of hydrophilic compounds that occurs via a series of reactions; that is, reductive dissolution of chalcopyrite by Na_2SO_3 (Equation (13)), oxidative dissolution of Cu_2S (Equation (14)), and hydrolysis/precipitation of Cu^{2+} and Fe^{3+} (Equations (15)–(18)).

$$2CuFeS_2 + 6Cu^{2+} + 3SO_3^{2-} + 6OH^- \rightarrow 4Cu_2S + 2Fe^{3+} + 3SO_4^{2-} + 3H_2O \tag{13}$$

$$Cu_2S \rightarrow 2Cu^{2+} + S + 4e^- \tag{14}$$

$$Cu^{2+} + H_2O \rightarrow CuO + 2H^+ \tag{15}$$

$$Cu^{2+} + 2H_2O \rightarrow Cu(OH)_2 + 2H^+ \tag{16}$$

$$Fe^{3+} + 2H_2O \rightarrow FeOOH + 3H^+ \tag{17}$$

$$2Fe^{3+} + 3SO_4^{2-} \rightarrow Fe_2(SO_4)_3 \tag{18}$$

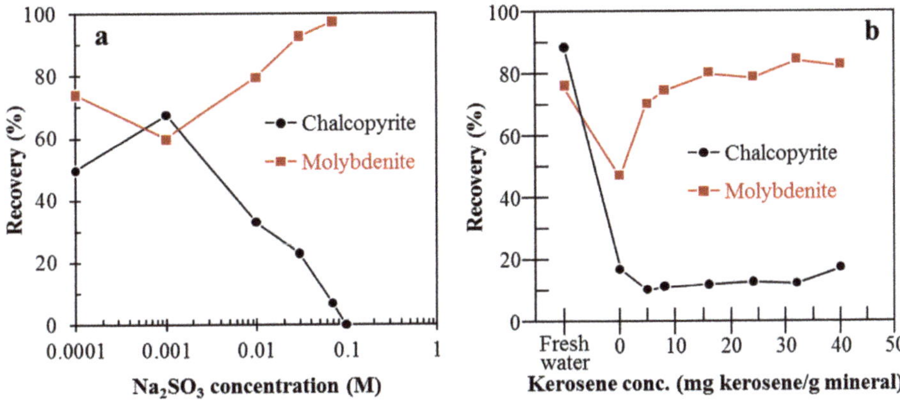

Figure 8. Recovery of chalcopyrite and molybdenite with varying concentrations of (**a**) Na_2SO_3 in TMD solution (mixture of TX15216 (alkyl mercaptan), MX-7017 (modified thionocarbamate), and diesel oil) [43] and (**b**) kerosene in $MgCl_2$ solution (reprinted with permission from Hirajima et al. [54], copyright (2016) Elsevier).

Meanwhile, the change in molybdenite surface after Na_2SO_3 treatment was almost negligible, indicating that the molybdenite surface remained hydrophobic. These results support the potential of Na_2SO_3 as a Cu depressant for Cu–Mo flotation separation. However, Na_2SO_3 does not always depress the floatability of chalcopyrite, but it can act as an activator depending on the state of the chalcopyrite surface. In the case of chalcopyrite, of which the surface is significantly covered with ferric oxyhydroxide, for example, the floatability of chalcopyrite is depressed to entrainment level (i.e., <10%) due to the presence of the hydrophilic nature of ferric oxyhydroxide on its surface. Meanwhile, the depressing effect of adsorbed ferric species is diminished when Na_2SO_3 is introduced because ferric species are reductively dissolved by SO_3^{2-} (Equations (19) and (20)), resulting in the exposure of iron-deficient chalcopyrite [45]. Therefore, the use of Na_2SO_3 as a Cu depressant should be carefully designed to avoid the failure of Cu–Mo separation.

$$SO_4^{2-} + H_2O + 2e^- \rightarrow SO_3^{2-} + 2OH^-, E^0 = -0.93 \text{ V} \tag{19}$$

$$Fe(OH)_3 + e^- \rightarrow Fe(OH)_2 + OH^-, E^0 = -0.56 \text{ V} \tag{20}$$

In arid and semiarid regions where the supply of fresh water is limited, the use of seawater in mineral-processing plants is a sustainable option. However, seawater, a concentrated solution of NaCl (around 0.6 M) with various secondary ions (e.g., 0.41 g/L Ca^{2+}, 1.28 g/L Mg^{2+}, 2.71 g/L SO_4^{2-}, and 0.11 g/L HCO_3^-), significantly changes the flotation behavior of molybdenite [46–49]. For example, the floatability of chalcopyrite is not affected by types of water sources (i.e., fresh and sea water) in the pH range of 7.5–11.5, but molybdenite is dramatically depressed in seawater when pH is higher than 9.5–10.0 [47,50–52]. This indicates that a significant amount of molybdenite is lost during rougher and cleaner flotation if the pulp pH is adjusted to 10–12 by using lime for depressing pyrite [47]. The primary detrimental effect on molybdenite floatability is due to the adsorption of magnesium hydroxy complexes ($Mg(OH)^+_{(aq)}$) and magnesium hydroxide precipitates ($Mg(OH)_{2(s)}$), which start forming at a pH above around 9.5 [50]. Qui et al. [53] analyzed the surface of molybdenite conditioned in seawater at pH 11 by using scanning electron microscopy (SEM), auger electron spectroscopy (AES), XPS, and ToF-SIMS, and confirmed that colloidal $Mg(OH)_2$ and crystallized $CaCO_3$ were deposited on the molybdenite surface. Meanwhile, Hirajima et al. [54] reported that $Mg(OH)_2$ depressed not only molybdenite but also chalcopyrite at pH > 9. Similarly, Nagaraj and Farinato [55] also reported that both molybdenite and chalcopyrite floatability was decreased in Mg^{2+}-containing solution at pH > 9.5 although the former was more strongly affected than the latter was. These indicate that seawater does not always show identical impact on Cu recovery, which changes depending on the type of used reagents and ore compositions [49]. In the case that both Cu and Mo were depressed by $Mg(OH)_2$, Hirajima and coworkers [54] proposed the utilization of kerosene emulsion for the flotation separation of chalcopyrite and molybdenite in a 0.01 M $MgCl_2$ solution (equivalent to about 243 mg/L Mg^{2+}). As shown in Figure 8b, the recoveries of chalcopyrite and molybdenite in 0.01 M $MgCl_2$ solution at pH 11 decreased from 90% to 18% and from 75% to 50%, respectively; however, molybdenite recovery was selectively improved with the addition of kerosene emulsion. Comparing the dynamic-force-microscopy (DFM) images of molybdenite surfaces conditioned in 0.01 M $MgCl_2$ at pH 11 with and without emulsified kerosene, the coverage of precipitate on molybdenite surface was apparently lower when kerosene was conditioned together. This is most likely because kerosene is a nonpolar oily collector that has strong affinity with molybdenite; thus, it is preferably adsorbed on the molybdenite surface, which limits the attachment of $Mg(OH)_2$. The study of Suyantara et al. [49], a follow-up study of Hirajima et al. [54], reported that kerosene emulsion could also be used for improving Cu–Mo flotation separation in artificial seawater. In the case that only molybdenite is depressed in seawater under alkaline conditions (i.e., pH 10–12), three options can be proposed: (1) the use of dispersants (e.g., sodium hexametaphosphate (SHMP)) to disperse seawater precipitates adsorbed on molybdenite surface [56,57]; (2) the removal of problematic ions (e.g., Ca^{2+} and Mg^{2+}) using Na_2CO_3 and CaO prior to rougher/cleaner flotation [58]; and (3) the operation of rougher/cleaner flotation at a pH lower than 9.5, where Ca^{2+}/Mg^{2+} precipitations are limited, with alternative pyrite depressants that can act at pH < 9.5 [47].

3.2.2. Organic Depressants

Along with the investigation on inorganic depressants, there have been many studies on organic depressants; for example, pseudo-glycolythiourea acid (PGA) [59], 2,3-disulfanylbutanedioic acid (DMSA) [60], disodium bis (carboxymethyl) trithiocarbonate (DBT) [61], chitosan [62], 4-amino-3-thioxo-3,4-dihydro-1,2,4-triazin-5(2H)-one (ATDT) [66], and acetic acid-[(hydrazinylthioxomethyl)thio]-sodium (AHS) [67]. Chen and coworkers [59] used PGA (Figure 9a) and investigated its effect on Cu–Mo separation. Under the optimized conditions, Mo recovery reached around 90% with a grade of 26% (original grade in feed = 0.4%) via one rougher (375 g/t Na_2SiO_3, 4000 g/t PGA, and 200 g/t kerosene), one scavenger (500 g/t PGA and 200 g/t kerosene), and two cleaners (62 g/t Na_2SiO_3, 1000 g/t, and 66 g/t kerosene), where Na_2SiO_3, PGA, and kerosene were used as a dispersant, Cu depressant, and Mo collector, respectively. Meanwhile, the recovery of Cu was around 0.7%, and its grade decreased from 18% (Cu grade in feed) to 10% (Cu grade in concentrate), indicating that PGA could be used as a Cu depressant in Cu–Mo flotation. To understand the mechanism of how PGA selectively depresses chalcopyrite, Chen et al. [59] conducted PGA adsorption tests with single minerals of chalcopyrite and molybdenite, and the results suggested that PGA was adsorbed on both minerals' surfaces, but the adsorption capacity of chalcopyrite was much larger compared to that of molybdenite. Infrared (IR) analyses of minerals after PGA adsorption tests indicated that the IR spectrum of PGA-adsorbed chalcopyrite was significantly changed with the appearance of new peaks (e.g., —COOH and —CS— of PGA), the disappearance of chalcopyrite peaks, and the great shift of SO_4^{2-} peaks. These changes in IR signatures were most likely caused by the chemisorption between S atoms in PGA, the most reactive site based on its largest electron density of highest occupied molecular orbital (HOMO), and Cu atoms on the chalcopyrite surface. On the other hand, the IR spectrum of molybdenite showed only small shifts of absorption peaks less than 3 cm^{-1}, which indicated that PGA is physically adsorbed on the surface of molybdenite. Even in the presence of butyl xanthate (BX), the most common collector for sulfide flotation, PGA could be adsorbed on chalcopyrite and molybdenite surfaces. The adsorption capacity of BX on both minerals' surfaces decreased as PGA concentration increased, which means that BX adsorption was limited due to the competitive adsorption between PGA and BX on chalcopyrite and molybdenite surfaces. Moreover, PGA has strong reducibility that can reduce dixanthogen molecules formed on chalcopyrite and molybdenite surfaces via the electrochemical reaction of BX. Fermi energies (E_F) of PGA, molybdenite, and chalcopyrite are −3.936, −4.138, and −5.433 eV, respectively. The direction of electron transition is always from a high- to a low-energy level; thus, PGA preferably gave the electrons to chalcopyrite than to molybdenite when the two minerals coexisted.

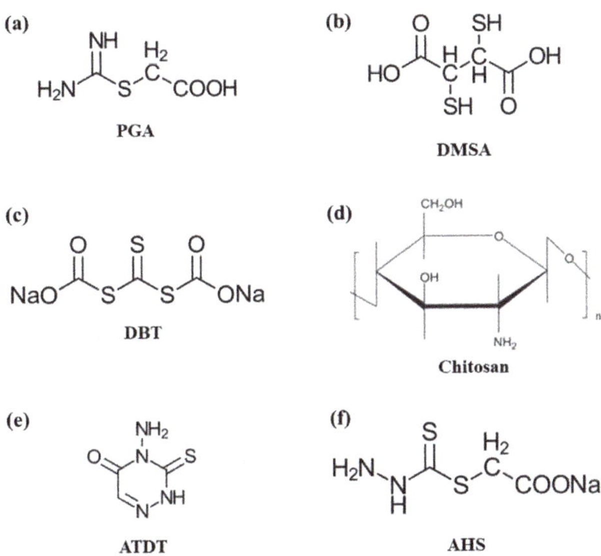

Figure 9. Molecular structure of organic depressants: (**a**) pseudo-glycolythiourea acid (PGA), (**b**) 2,3-disulfanylbutanedioic acid (DMSA), (**c**) disodium bis (carboxymethyl) trithiocarbonate (DBT), (**d**) chitosan (reprinted with permission from Crini and Badot [65], copyright (2008) Elsevier), (**e**) 4-amino-3-thioxo-3,4-dihydro-1,2,4-triazin-5(2H)-one (ATDT), and (**f**) acetic acid-[(hydrazinylthioxomethyl)thio]-sodium (AHS).

A novel and selective depressant named DMSA was first used in the flotation of Cu–Mo sulfides by Li and coworkers [60]. As illustrated in Figure 9b, DMSA contains two carboxyl and two sulfhydryl groups. Adsorption tests showed that DMSA can be adsorbed more intensively on chalcopyrite than molybdenite can; that is, the adsorption capacities of DMSA on chalcopyrite and molybdenite were 0.29 and 0.05 mg/g, respectively. According to the calculation based on frontier-molecular-orbital (FMO) theory, the largest electron density of HOMO is located on two S atoms, both of which actively interact with minerals. Additionally, it is interesting to note that the Fermi energies of DMSA, molybdenite, and chalcopyrite are −5.573, −4.353, and −6.110 eV, respectively, which indicate that the electron transfer between DMSA and chalcopyrite, an essential process for the adsorption of organic with a strong reducibility, occurs more easily than that between DMSA and molybdenite, consistent with the adsorption results. A similar result was obtained from the study of Yin et al. [61] who used DBT as a depressant (Figure 9c). The IR spectrum of molybdenite after DBT treatment showed that the characteristic bands of DBT were not observed; however, the IR spectrum of DBT-treated chalcopyrite showed the appearance of new peaks of C=S stretching, C—O stretching, CH_2 wagging, CH_2 scissoring, and C—C stretching, not observed in the IR spectrum of chalcopyrite without DBT treatment, indicating that DBT was notably adsorbed on the chalcopyrite surface. The DBT-treated chalcopyrite was further analyzed by XPS, and the C 1s and S 2p spectra also proved the presence of DBT on its surface. The S $2p_{3/2}$ peaks of DBT adsorbed on the chalcopyrite surface were shifted to higher binding energies than those of DBT, while the binding energies of Cu $2p_{3/2}$ and Cu $2p_{1/2}$ were shifted to lower positions after DBT was adsorbed. From these results, Yin and coworkers [61] concluded that S atoms in DBT (i.e., the electron donor) might interact with the Cu atoms in chalcopyrite (i.e., the electron acceptor). Due to the selective and strong affinity of DMSA and DBT with chalcopyrite, those organic compounds could be successfully used as a Cu depressant in Cu–Mo flotation.

As such, organic depressants that contain S atom(s) as a constituent in their molecules, such as PGA, DMSA, and DBT, have the strong ability to selectively depress chalcopyrite in Cu–Mo concentrates. Another attempt of using chitosan (Figure 9d) has also been made for the same purpose [62–64]. Chitosan, a natural biodegradable and nontoxic polyaminosaccharide, contains an amino functional group (—NH_2) known to have strong affinity with Cu^{2+}, which implies that it can specifically interact with Cu-bearing minerals. Li and coworkers [62] used chitosan as a Cu depressant in Cu–Mo flotation. According to the result of single-mineral flotation, chitosan depressed both chalcopyrite and molybdenite, but it was effective in selectively depressing chalcopyrite when both minerals existed together in the pulp. This result could be explained by the competitive adsorption of chitosan between the two minerals, proved by adsorption isotherm tests, which showed that the adsorption density of chitosan on chalcopyrite was greater than that on molybdenite. As a result, chitosan could succeed in the selective flotation of molybdenite from Cu–Mo concentrates, that is, the recoveries of molybdenite and chalcopyrite in the froth product were around 70% and 24%, respectively. Furthermore, organic compounds that contained both amine- and thione-functional groups, such as ATDT (Figure 9e) and AHS (Figure 9f), were reported to be effective in depressing chalcopyrite due to bidentate coordination to $Cu^{I/II}$ via S and N atoms to form a stable five-membered chelating ring (Figure 10) [66,67]. The selectivity index (SI) of Mo/Cu in the absence of a depressant was around 1.44, but considerably improved to 2.98 with ATDT and to 5.77 with AHS.

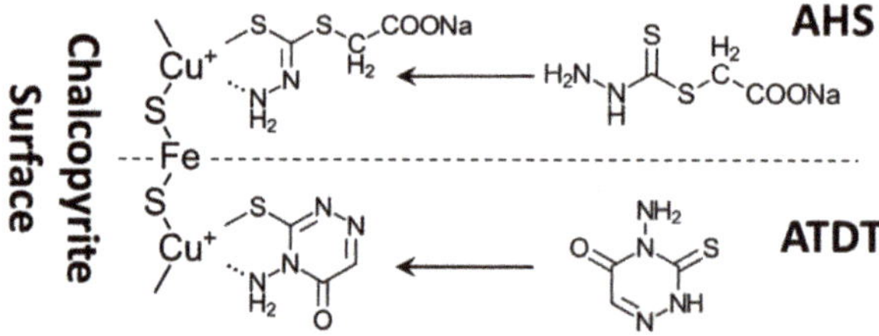

Figure 10. Proposed adsorption models of AHS and ATDT on chalcopyrite surface.

3.3. Oxidation Treatments for Depressing Cu Minerals

3.3.1. Ozone Oxidation

In the early 1990s, the application of ozone (O_3) for the depression of Cu minerals (e.g., chalcopyrite and chalcocite) and the selective recovery of molybdenite from Cu–Mo bulk concentrates was studied [68,69]. Ozone, a strong oxidant, as illustrated in Equations (21) and (22), is widely used for water treatment due to its exceptional ability to oxidize organic compounds and to disinfect bacteria [70,71].

$$O_3 + 6H^+ + 6e^- \rightarrow 3H_2O, \; E^0 = 1.51 \text{ V} \tag{21}$$

$$O_3 + 2H^+ + 2e^- \rightarrow O_2 + H_2O, \; E^0 = 2.07 \text{ V} \tag{22}$$

Moreover, the decomposition of ozone in water produces hydroxyl radicals ($HO^\bullet$) (Equation (23)), which are among the most reactive free radicals and among the strongest oxidants (Equation (24)) [71,72].

$$O_3 + H_2O \rightarrow 2HO^\bullet + O_2 \tag{23}$$

$$HO^\bullet + H^+ + e^- \rightarrow H_2O, \; E^0 = 2.33 \text{ V} \tag{24}$$

In the typical flotation process of Cu–Mo ores (Figure 2), the flotation separation of Cu and Mo from Cu–Mo bulk concentrates is achieved by using NaHS, which desorbs the adsorbed xanthate on the chalcopyrite surface, that is, chalcopyrite becomes unrecoverable as a froth product while molybdenite is selectively floated and recovered. Although xanthates are known as weak collectors of molybdenite, it can also be adsorbed on the surface of molybdenite [73]. However, the molybdenite surface is naturally hydrophobic, so even in the absence of xanthate on its surface, molybdenite is still floatable with nonpolar oily collectors (e.g., diesel oil and kerosene), which makes the desorption of xanthate by NaHS strongly affect the floatability of chalcopyrite rather than that of molybdenite. Similar to NaHS, the utilization of ozone, having strong potential to oxidize organic compounds, can improve Cu–Mo flotation separation by destroying xanthate adsorbed onto the mineral surface. Ye et al. [69] utilized ozone as a conditioning process for Cu–Mo separation, and reported that ozone conditioning oxidized all tested minerals (e.g., chalcocite, chalcopyrite, and molybdenite), leading to their floatability lowering. However, Cu minerals are more sensitive to ozone compared to molybdenite, so the selective depression of Cu minerals is possible by employing ozone-based oxidation treatment. For example, a flotation circuit that consists of a rougher flotation with ozone conditioning for 2 min, followed by a cleaner flotation after 3 min of ozone treatment of low-grade Cu–Mo concentrate (0.25% Mo), can produce a Mo concentrate of which grade and recovery are 26% and 82.5%, respectively [69]. After rougher flotation, 10.5% of molybdenite still remained in rougher tailing, which was further conditioned with ozone for 1 min, followed by scavenger flotation. The distributions of Mo in scavenger concentrate and cleaner tailings are 9.4% and 7.0%, respectively, indicating that overall Mo recovery would be expected to reach 98.9% when they were reintroduced to the process (denoted as dotted lines in Figure 11).

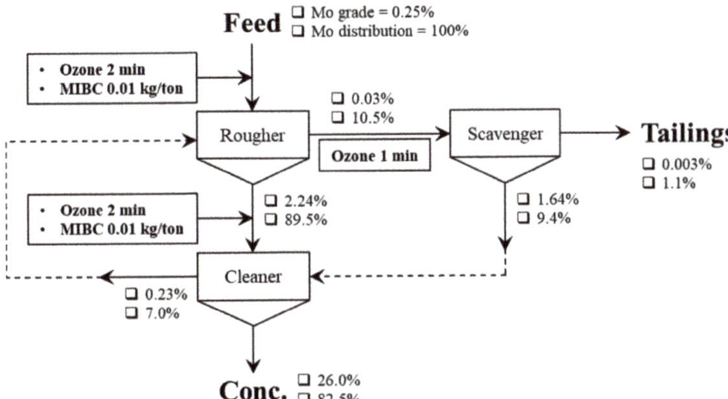

Figure 11. Flowsheet for selective flotation of molybdenite from Cu–Mo concentrates using three-stage ozone conditioning. Dotted lines represent possible recycle streams (reprinted with permission from Ye et al. [69], copyright (1990) Springer Nature).

3.3.2. Plasma Oxidation

Similarly, Hirajima et al. [74] examined oxidation treatment using plasma as a conditioning process for the selective flotation of chalcopyrite and molybdenite. After 10 min treatment of chalcopyrite with plasma at 10 W, AFM images showed that the chalcopyrite surface became rough (3.979 nm) compared to the one without plasma treatment (1.280 nm), and there was no significant effect of washing on the change in surface roughness (Figure 12). In the case of molybdenite, its surface roughness increased from 3.797 to 5.176 nm after plasma treatment (10 W) for 10 min, but it became smooth when washing followed; that is, the washing of plasma-treated molybdenite for 30, 60, and 120 min decreased its surface roughness to 4.263, 3.484, and 3.289 nm, respectively. The presence of reaction products on the surfaces of chalcopyrite and molybdenite after plasma treatment significantly affected their floatability. As shown in Figure 13a, the recovery of untreated chalcopyrite and molybdenite was around 90%, but dramatically decreased to 10–30% after plasma treatment. Washing following could improve the floatability of molybdenite treated with plasma, whereas chalcopyrite recovery was kept constant regardless of washing. To clarify the flotation behaviors of chalcopyrite and molybdenite with plasma treatment followed by washing, Hirajima et al. [74] conducted XPS analysis of plasma-treated minerals with various washing times. The result of XPS analysis of plasma-treated molybdenite showed that it was covered with molybdenum(VI) oxide (MoO_3); however, the peaks of MoO_3 in Mo 3d and O 1s spectra were lowered after washing. In the case of plasma-treated chalcopyrite, its surface was covered with goethite (FeOOH) and iron(III) sulfate ($Fe_2(SO_4)_3$), both of which are oxidation products of chalcopyrite having hydrophilic properties. After washing, the peaks of iron(III) sulfate disappeared, while goethite still remained on the surface of chalcopyrite. This is the reason why the floatability of plasma-treated chalcopyrite did not increase even after washing for 120 min. Washing cannot improve molybdenite floatability when both minerals are treated together by plasma. This might be due to iron ions (e.g., Fe^{2+} and/or Fe^{3+}) being released from chalcopyrite that are precipitated on the surface of not only chalcopyrite but also molybdenite, reducing the hydrophobicity of both minerals. To overcome this limited recovery of molybdenite, Hirajima and coworkers [74] investigated the addition of kerosene, a commonly used reagent for Mo flotation. As a result of the addition of 25 μL emulsified kerosene, molybdenite recovery increased from around 40% to 80%, while it had a negligible effect on chalcopyrite recovery, increasing from 20% to 30%.

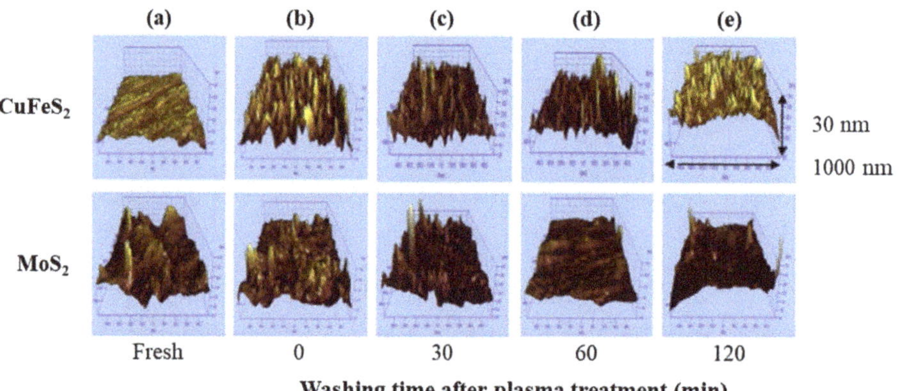

Figure 12. AFM images of chalcopyrite and molybdenite: (**a**) untreated samples and one treated with plasma at 10 W for 10 min followed by washing by pH 9 solution for (**b**) 0, (**c**) 30, (**d**) 60, and (**e**) 120 min (reprinted with permission from Hirajima et al. [74], copyright (2014) Elsevier).

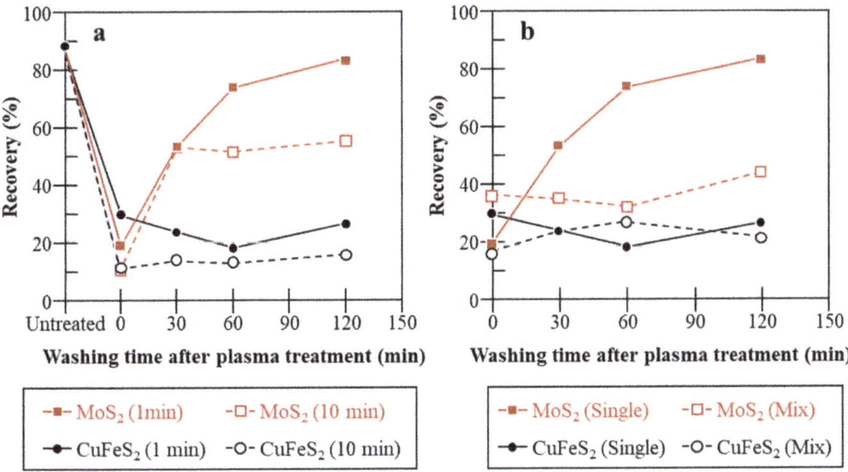

Figure 13. Flotation recovery of chalcopyrite and molybdenite treated (**a**) with and without 10 W plasma for 1 or 10 min followed by washing (single system), and (**b**) with 10 W plasma for 1 min followed by washing (single and mixed system) (reprinted with permission from Hirajima et al. [74], copyright (2014) Elsevier).

3.3.3. H_2O_2 Oxidation

Although plasma treatment for Cu–Mo ores prior to flotation was effective in improving the separation of chalcopyrite and molybdenite, its application on an industrial scale remains difficult to realize [75]. Because of this limitation, Hirajima and coworkers [13] examined another oxidation process using hydrogen peroxide (H_2O_2), which acts as an oxidant under acidic conditions (Equation (25)), but behaves as a reductant under alkaline conditions (Equation (26)).

$$H_2O_2 + 2H^+ + 2e^- \rightarrow 2H_2O, E^0 = 1.78 \text{ V} \tag{25}$$

$$H_2O_2 \rightarrow 2H^+ + O_2 + 2e^-, E^0 = -0.68 \text{ V} \tag{26}$$

Moreover, H_2O_2 can produce hydroxyl and hydroperoxyl radicals ($HO^\bullet$ and $HOO^\bullet$, respectively) acting as powerful oxidizing agents in the presence of iron ions (e.g., Fe^{2+} and Fe^{3+}) via Fenton (Fe^{2+}/H_2O_2) and Fenton-like (Fe^{3+}/H_2O_2) reactions, as shown in the following equations:

$$Fe^{2+} + H_2O_2 \rightarrow Fe^{3+} + HO^\bullet + OH^-, \tag{27}$$

$$Fe^{3+} + H_2O_2 \rightarrow Fe^{2+} + HOO^\bullet + H^+. \tag{28}$$

After 10 min treatment of chalcopyrite using 0.1% H_2O_2, the floatability of chalcopyrite decreased from 85% to 19%, most likely due to the formation of hydrophilic oxidation products (e.g., CuO, Cu(OH)$_2$, FeOOH, and Fe$_2$(SO$_4$)$_3$) on the chalcopyrite surface. Meanwhile, the floatability of molybdenite was not affected by H_2O_2 treatment (the recoveries of untreated and H_2O_2-treated molybdenite were 73% and 79%, respectively) because the oxidation products of molybdenite (e.g., MoO$_2$ and MoO$_3$) are soluble under alkaline conditions. Hirajima et al. [13] also compared the effectiveness of oxidation treatments using ozone and H_2O_2 on Cu–Mo flotation separation, and it was confirmed that H_2O_2 treatment showed better separation efficiency compared to that of ozone treatment because ozone reduced the recovery of not only chalcopyrite (from 85% to 28%), but also molybdenite (from 73% to 53%). Although effective, the application of H_2O_2 treatment to Cu–Mo bulk concentrates with high pulp density (around 50%) has some drawbacks compared to the NaHS method: (1) molybdenite recovery by H_2O_2-based flotation is lower than that of conventional flotation process using NaHS; (2) it requires a prolonged treatment of around 4.5 h to effectively depress the floatability of chalcopyrite, while 10 min is enough for the NaHS method; and (3) it uses a high concentration of H_2O_2 (i.e., 2%), which makes H_2O_2-based flotation costly [75]. To overcome these limitations of H_2O_2 oxidation treatment, Suyantara and coworkers [75] investigated the simultaneous use of H_2O_2 and $FeSO_4$, which can stimulate a Fenton-like reaction. In an alkaline solution, $FeSO_4$ releases Fe^{2+}, which is then transformed into iron oxyhydroxide (goethite, α-FeOOH; lepidocrocite, γ-FeOOH) and/or ferrihydrite, where a Fenton-like reaction occurs in the presence of H_2O_2 as shown in the following equations [76]:

$$\equiv Fe^{III}-OH + H_2O_2 \rightarrow (H_2O_2)_s \tag{29}$$

$$(H_2O_2)_s \rightarrow \equiv Fe^{II} + H_2O + HOO^\bullet \tag{30}$$

$$\equiv Fe^{II} + H_2O_2 \rightarrow \equiv Fe^{III}-OH + HO^\bullet \tag{31}$$

$$HOO^\bullet \rightarrow H^+ + O_2^{\bullet-} \tag{32}$$

$$\equiv Fe^{III}-OH + HOO^\bullet/O_2^{\bullet-} \rightarrow \equiv Fe^{II} + H_2O/OH^- + O_2. \tag{33}$$

According to Li et al. [77], who evaluated the decomposition rate of H_2O_2 in the absence and presence of FeOOH, it was confirmed that the addition of FeOOH significantly enhanced H_2O_2 decomposition compared to the one without FeOOH. Suyantara et al. [75], who used H_2O_2 with $FeSO_4$ for the oxidation treatment of Cu–Mo bulk concentrates prior to flotation, confirmed that the addition of $FeSO_4$ could reduce the amount of added H_2O_2 from 2.0% to 0.5% and treatment time from 4.5 h to 5 min. Due to these positive effects of $FeSO_4$ addition to H_2O_2 oxidation treatment, the process of using H_2O_2 and $FeSO_4$ is predicted to reduce operating costs, that is, the costs for NaHS, H_2O_2, and $H_2O_2/FeSO_4$ methods were calculated to approximately 20, 100, and 15 USD/t, respectively [75].

3.3.4. Electrolysis Oxidation

The electrical resistivities of chalcopyrite and molybdenite are 234 Ω and 1.2–1.5 MΩ, respectively, which means that chalcopyrite is more electrochemically active than molybdenite is [78]. From the difference in the minerals' electrical resistivity, Miki et al. [78] attempted to apply electrolysis, a technique that applies a fixed potential in which mineral undergoes oxidation process(es) to selectively render the chalcopyrite surface hydrophilic. Anodic polarization results at an applied potential of 1.2 V showed that a high current density of chalcopyrite electrodes (around 0.4 mA/m^2) was observed, indicating that the oxidation of chalcopyrite was actively progressed, as shown in the following equation:

$$CuFeS_2 + 8H_2O \rightarrow Cu^{2+} + Fe^{2+} + 2SO_4^{2-} + 16H^+ + 16e^-. \tag{34}$$

On the other hand, the current density of molybdenite electrodes was around 0.05–0.15 mA/m^2, substantially lower compared to that of chalcopyrite. This most likely resulted from the high resistivity of molybdenite, which makes it difficult to be oxidized. After treating chalcopyrite and molybdenite electrodes by electrolysis oxidation at 1.2 V for 800 s, the contact angle of chalcopyrite changed from 71.7° to 42.4°, whereas there was almost no effect on the contact angle of molybdenite, changing from 68.6° to 67.0°. The contact-angle results indicate that electrolysis treatment could selectively convert the wettability of chalcopyrite from hydrophobic to hydrophilic. Although flotation tests of electrolysis-treated chalcopyrite/molybdenite were not conducted, electrolysis oxidation is in a promising state as pretreatment for improving the selective flotation of molybdenite from Cu–Mo concentrates.

3.4. Microencapsulation Techniques for Depressing Cu Minerals

As discussed in Section 3.3.4, chalcopyrite and molybdenite have different electrical resistivities, which means that the depression techniques involved in electrochemical processes have the potential to be applied to Cu–Mo flotation separation. For example, the authors studied carrier microencapsulation (CME), a technique that creates metal-oxyhydroxide coatings on the surfaces of semiconducting minerals such as pyrite and arsenopyrite (FeAsS) [79–84]. In CME, metal ions (e.g., Al^{3+}, Fe^{3+}, Si^{4+}, and Ti^{4+}) and the organic carrier (e.g., catechol, 1,2-dihydroxybenzene, and $C_6H_4(OH)_2$) are used, and they produce various forms of metal–catecholate complexes (e.g., $[Al(cat)_n]^{3-2n}$, $[Fe(cat)_n]^{3-2n}$, $[Si(cat)_3]^{2-}$ and $[Ti(cat)_3]^{2-}$, where n is 1–3). These complexes undergo oxidative decomposition only on the surface of semiconducting minerals such as pyrite and arsenopyrite (FeAsS), and metal ions released from the complexes are then precipitated as metal oxyhydroxides (Figure 14). Similar to CME, other microencapsulation techniques involved in electrochemical process(es) are most likely able to selectively form hydrophilic coatings on the surface of chalcopyrite having relatively low electrical resistivity, while the molybdenite surface is not covered with the coatings because it is hard for electrochemical reactions to occur on its surface. Although there is no study on this topic, it is of importance to develop environmentally friendly processes for Cu–Mo flotation separation.

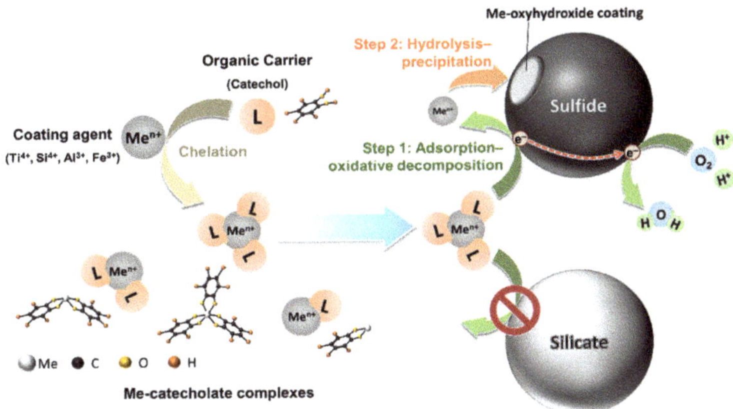

Figure 14. Schematic diagram of carrier microencapsulation (reprinted with permission from Park et al. [83], copyright (2019) Elsevier).

4. Summary

Porphyry copper deposits are one of the most important sources of copper and molybdenum. The separation/recovery of copper and molybdenum from these deposits is typically processed by a flotation series, that is, bulk flotation to remove gangue minerals and produce Cu–Mo concentrates that are then processed by Cu–Mo flotation to separate them. In Cu–Mo flotation, various types of Cu depressants (e.g., NaHS, Na_2S, Nokes reagent, and NaCN) have been adopted for selective depression of Cu minerals. However, these reagents are potentially dangerous due to the possibility of emitting toxic and deadly gases, such as H_2S and HCN, when operating conditions are not properly controlled. To avoid accidents caused by the use of conventional Cu depressants, there are many studies on the utilization of environmentally friendly depressants for molybdenite or chalcopyrite, and oxidation treatments for chalcopyrite.

For the depression of molybdenite, various organic compounds, such as dextrin, lignosulfonate, O-carboxymethyl chitosan, carboxymethyl cellulose, and humic acid, were adopted. Molybdenite is well-known to have strong hydrophobicity, so those organic compounds are favorably and selectively adsorbed on the molybdenite surface via hydrophobic interactions. The utilization of Mo depressants is effective for the separation of Cu and Mo minerals, but there is a serious drawback, lowering the purity of froth products. This is because the content of molybdenite in PCDs is substantially lower than that of chalcopyrite, so the strategy of recovering chalcopyrite with the depression of molybdenite can cause the mechanical entrainment of molybdenite within a large volume of Cu concentrate, which lowers the grade of Cu concentrate and leads to appreciable loss of molybdenite. Another option for Cu–Mo flotation separation is to depress Cu minerals while recovering molybdenite. For this, alternative organic/inorganic depressants were extensively examined. Moreover, oxidation treatments using ozone, plasma, H_2O_2, and electrolysis to destroy adsorbed xanthate and/or to create hydrophilic coatings on the surface of chalcopyrite were studied. The obtained results are promising for Cu–Mo separation; however, there still remain important topics on the optimization and scaling-up of the process that are necessary for the application of newly developed depression techniques to actual Cu–Mo flotation separation.

Author Contributions: Conceptualization, I.P., S.H., S.J., M.I., and N.H.; writing—original-draft preparation, I.P.; writing—review and editing, I.P., S.H., S.J., M.I., and N.H.; visualization, I.P. All authors have read and agreed to the published version of the manuscript.

Funding: This research was financially supported by Japan Oil, Gas, and Metals National Corporation (JOGMEC).

Conflicts of Interest: The authors declare no conflict of interest.

References

1. John, D.A.; Ayuso, R.A.; Barton, M.D.; Blakely, R.J.; Bodnar, R.J.; Dilles, J.H.; Gray, F.; Graybeal, F.T.; Mars, J.C.; McPhee, D.K.; et al. *Porphyry Copper Deposit Model: Chapter B of Mineral Deposit Models for Resource Assessment*; Scientific Investigations Report 2010–5070–B; US Geological Survey: Menlo Park, CA, USA, 2010; 169p.

2. John, D.A.; Taylor, R.D. By-products of porphyry copper and molybdenum deposits. *Rev. Econ. Geol.* **2016**, *18*, 137–164.

3. Lee, C.-T.A.; Tang, M. How to make porphyry copper deposits. *Earth Planet. Sci. Lett.* **2020**, *529*, 115868. [CrossRef]

4. Seedorff, E.; Dilles, J.H.; Proffett, J.M., Jr.; Einaudi, M.T.; Zurcher, L.; Stavast, W.J.A.; Johnson, D.A.; Barton, M.D. Porphyry deposits: Characteristics and origin of hypogene features. *Econ. Geol.* **2005**, *100*, 251–298.

5. Sinclair, W.D. Porphyry deposits. In *Mineral Deposits of Canada: A Synthesis of Major Deposit-Types, District Metallogeny, the Evolution of Geological Provinces, and Exploration Methods, Geological Association of Canada, Mineral Deposits Division*; Goodfellow, W.D., Ed.; Special Publication: London, UK, 2007; Volume 5, pp. 223–243.

6. Bulatovic, S.M. 12–Flotation of Copper Sulfide Ores. In *Handbook of Flotation Reagents*; Bulatovic, S.M., Ed.; Elsevier: Amsterdam, The Netherlands, 2007; pp. 235–293, ISBN 978-0-444-53029-5.

7. Amelunxen, P.; Schmitz, C.; Hill, L.; Goodweiler, N.; Andres, J. Molybdenum. In *SME Mineral Processing & Extractive Metallurgy Handbook*; Dunne, R.C., Komar Kawatra, S., Young, C.A., Eds.; Society for Mining, Metallurgy & Exploration (SME): Littleton, CO, USA, 2019; Volume 2, pp. 1891–1916.

8. Liu, G.; Lu, Y.; Zhong, H.; Cao, Z.; Xu, Z. A novel approach for preferential flotation recovery of molybdenite from a porphyry copper-molybdenum ore. *Miner. Eng.* **2012**, *36*, 37–44. [CrossRef]

9. Peterson, J.A.; Saran, M.S.; Wisnouskas, J.S. Differential Flotation Reagent for Molybdenum Separation. U.S. Patent No. 4,575,419, 11 March 1986.

10. Yin, Z.; Sun, W.; Hu, Y.; Zhai, J.; Qingjun, G. Evaluation of the replacement of NaCN with depressant mixtures in the separation of copper-molybdenum sulphide ore by flotation. *Sep. Purif. Technol.* **2017**, *173*, 9–16. [CrossRef]

11. Sutherland, K.L.; Wark, I.W. *Principles of Flotation*, 2nd ed.; Australasian Inst. Mining & Metallurgy: Melbourne, Australia, 1955.

12. Mu, Y.; Peng, Y.; Lauten, R.A. The depression of pyrite in selective flotation by different reagent systems—A literature review. *Miner. Eng.* **2016**, *96–97*, 143–156. [CrossRef]

13. Hirajima, T.; Miki, H.; Suyantara, G.P.W.; Matsuoka, H.; Elmahdy, A.M.; Sasaki, K.; Imaizumi, Y.; Kuroiwa, S. Selective flotation of chalcopyrite and molybdenite with H_2O_2 oxidation. *Miner. Eng.* **2017**, *100*, 83–92. [CrossRef]

14. Napier-Munn, T.; Wills, B.A. *Wills' Mineral Processing Technology: An Introduction to the Practical Aspects of Ore Treatment and Mineral Recovery*, 7th ed.; Butterworth-Heinemann: Oxford, UK, 2006.

15. Zhao, Q.; Liu, W.; Wei, D.; Wang, W.; Cui, B.; Liu, W. Effect of copper ions on the flotation separation of chalcopyrite and molybdenite using sodium sulfide as a depressant. *Miner. Eng.* **2018**, *115*, 44–52. [CrossRef]

16. Guo, B.; Peng, Y.; Espinosa-Gomez, R. Cyanide chemistry and its effect on mineral flotation. *Miner. Eng.* **2014**, *66–68*, 25–32. [CrossRef]

17. Sinozaki, H.; Hara, R.; Mitsukuri, S. The vapour pressures of hydrogen cyanide. *Bull. Chem. Soc. Jpn.* **1926**, *1*, 59–61. [CrossRef]

18. Stull, D.R. Vapor pressure of pure substances. Organic and inorganic compounds. *Ind. Eng. Chem.* **1947**, *39*, 517–540. [CrossRef]

19. Milby, T.H.; Baselt, R.C. Hydrogen sulfide poisoning: Clarification of some controversial issues. *Am. J. Ind. Med.* **1999**, *35*, 192–195. [CrossRef]

20. National Research Council. *Acute Exposure Guideline Levels for Selected Airborne Chemicals: Volume 2*; The National Academies Press: Washington, DC, USA, 2002. [CrossRef]

21. Johnston, A.; Meadows, D.G.; Cappuccitti, F. Copper mineral processing. In *SME Mineral Processing & Extractive Metallurgy Handbook*; Dunne, R.C., Komar Kawatra, S., Young, C.A., Eds.; Society for Mining, Metallurgy & Exploration (SME): Littleton, CO, USA, 2019; Volume 2, pp. 1615–1642.

22. Hernlund, R.W. Extraction of molybdenite from copper flotation products. *Q. J. Colo. Sch. Mines* **1961**, *56*, 177–196.

23. Blanco, M.; Coello, J.; Iturriaga, H.; Maspoch, S.; González Bañó, R. On-line monitoring of starch enzymatic hydrolysis by near-infrared spectroscopy. *Analyst* **2000**, *125*, 749–752. [CrossRef]

24. Wie, J.M.; Fuerstenau, D.W. The effect of dextrin on surface properties and the flotation of molybdenite. *Int. J. Miner. Process.* **1974**, *1*, 17–32. [CrossRef]

25. Beaussart, A.; Mierczynska-Vasilev, A.; Beattie, D.A. Adsorption of dextrin on hydrophobic minerals. *Langmuir* **2009**, *25*, 9913–9921. [CrossRef]

26. Jorjani, E.; Barkhordari, H.R.; Tayebi Khorami, M.; Fazeli, A. Effects of aluminosilicate minerals on copper-molybdenum flotation from Sarcheshmeh porphyry ores. *Miner. Eng.* **2011**, *24*, 754–759. [CrossRef]

27. Ansari, A.; Pawlik, M. Floatability of chalcopyrite and molybdenite in the presence of lignosulfonates. Part I. Adsorption studies. *Miner. Eng.* **2007**, *20*, 600–608. [CrossRef]

28. Ansari, A.; Pawlik, M. Floatability of chalcopyrite and molybdenite in the presence of lignosulfonates. Part II. Hallimond tube flotation. *Miner. Eng.* **2007**, *20*, 609–616. [CrossRef]

29. Buza, T.B.; Hiscox, T.O.; Kuhn, M.C. Use of lignin sulphonate as moly depressant boosts recovery at Twin Buttes. *Eng. Min. J.* **1975**, *176*, 87–91.

30. Mathieu, G.; Bruce, R. Getting the talc out of molybdenite ores. *Can. Min. J.* **1974**, *95*, 75–77.

31. Yuan, D.; Cadien, K.; Liu, Q.; Zeng, H. Separation of talc and molybdenite: Challenges and opportunities. *Miner. Eng.* **2019**, *143*, 105923. [CrossRef]

32. Yuan, D.; Cadien, K.; Liu, Q.; Zeng, H. Flotation separation of Cu-Mo sulfides by O-Carboxymethyl chitosan. *Miner. Eng.* **2019**, *134*, 202–205. [CrossRef]

33. Yuan, D.; Cadien, K.; Liu, Q.; Zeng, H. Adsorption characteristics and mechanisms of O-Carboxymethyl chitosan on chalcopyrite and molybdenite. *J. Colloid Interface Sci.* **2019**, *552*, 659–670. [CrossRef] [PubMed]

34. Yuan, D.; Cadien, K.; Liu, Q.; Zeng, H. Selective separation of copper-molybdenum sulfides using humic acids. *Miner. Eng.* **2019**, *133*, 43–46. [CrossRef]

35. Kor, M.; Korczyk, P.M.; Addai-Mensah, J.; Krasowska, M.; Beattie, D.A. Carboxymethylcellulose adsorption on molybdenite: The effect of electrolyte composition on adsorption, bubble–surface collisions, and flotation. *Langmuir* **2014**, *30*, 11975–11984. [CrossRef]

36. Mierczynska-Vasilev, A.; Beattie, D.A. Adsorption of tailored carboxymethyl cellulose polymers on talc and chalcopyrite: Correlation between coverage, wettability, and flotation. *Miner. Eng.* **2010**, *23*, 985–993. [CrossRef]

37. Qui, X.; Yang, H.; Chen, G.; Luo, W. An Alternative Depressant of Chalcopyrite in Cu–Mo Differential Flotation and Its Interaction Mechanism. *Minerals* **2019**, *9*, 1.

38. Bulatovic, S.M. Use of organic polymers in the flotation of polymetallic ores: A review. *Miner. Eng.* **1999**, *12*, 341–354. [CrossRef]

39. Grano, S.; Ralston, J.; Smart, R.S.C. Influence of electrochemical environment on the flotation behavior of Mt. Isa copper and lead-zinc ore. *Int. J. Miner. Process.* **1990**, *30*, 69–97. [CrossRef]

40. Heyes, G.W.; Trahar, W.J. The natural flotability of chalcopyrite. *Int. J. Miner. Process.* **1977**, *4*, 317–344. [CrossRef]

41. Ralston, J. Eh and its consequences in sulphide mineral flotation. *Miner. Eng.* **1991**, *4*, 859–878. [CrossRef]

42. Richardson, P.E.; Walker, G.W. The flotation of chalcocite, bornite, chalcopyrite, and pyrite in an electrochemical-flotation cell. In Proceedings of the XV International Mineral Processing Congress, Cannes, France, 2–9 June 1985; pp. 198–210.

43. Guy, P.J.; Trahar, W.J. The effects of oxidation and mineral interaction on sulphide flotation. In *Flotation of Sulphide Minerals*; Forrsberg, K.S.E., Ed.; Elsevier: Amsterdam, The Netherlands, 1985; pp. 91–110.

44. Miki, H.; Hirajima, T.; Muta, Y.; Suyantara, G.P.W.; Sasaki, K. Effect of sodium sulfite on floatability of chalcopyrite and molybdenite. *Minerals* **2018**, *8*, 172. [CrossRef]

45. Grano, S.R.; Sollaart, M.; Skinner, W.; Prestidge, C.A.; Ralston, J. Surface modification in the chalcopyrite-sulphite ion system. I. collectorless flotation, XPS and dissolution study. *Int. J. Miner. Process.* **1997**, *50*, 1–26. [CrossRef]

46. Castro, S.; Laskowski, J.S. Froth flotation in saline water. *KONA Powder Part. J.* **2011**, *29*, 4–15. [CrossRef]

47. Castro, S. Challenges in flotation of Cu–Co sulfide ores in sea water. In *The First International Symposium on Water in Mineral Processing*; Drelich, J., Ed.; Society for Mining, Metallurgy & Exploration (SME): Littleton, CO, USA, 2012; pp. 29–40.

48. Laskowski, J.S.; Castro, S.; Ramos, O. Effect of seawater main components on frothability in the flotation of Cu-Mo sulfide ore. *Physicochem. Probl. Miner. Process.* **2013**, *50*, 17–29.

49. Suyantara, G.P.W.; Hirajima, T.; Miki, H.; Sasaki, K. Floatability of molybdenite and chalcopyrite in artificial seawater. *Miner. Eng.* **2018**, *115*, 117–130. [CrossRef]

50. Castro, S.; Rioseco, P.; Laskowski, J.S. Depression of molybdenite in sea water. In Proceedings of the XXVI International Mineral Processing Congress, New Delhi, India, 24–28 September 2012; pp. 737–752.

51. Laskowski, J.S.; Castro, S. Flotation in concentrated electrolyte solutions. *Int. J. Miner. Process.* **2015**, *144*, 50–55. [CrossRef]

52. Ramos, O.; Castro, S.; Laskowski, J.S. Copper-molybdenum ores flotation in sea water: Floatability and frothability. *Miner. Eng.* **2013**, *53*, 108–112. [CrossRef]

53. Qui, Z.; Liu, G.; Liu, Q.; Zhong, H. Understanding the roles of high salinity in inhibiting the molybdenite flotation. *Colloids Surf. A Physicochem. Eng. Asp.* **2016**, *509*, 123–129.

54. Hirajima, T.; Suyantara, G.P.W.; Ichikawa, O.; Elmahdy, A.M.; Miki, H.; Sasaki, K. Effect of Mg^{2+} and Ca^{2+} as divalent seawater cations on the floatability of molybdenite and chalcopyrite. *Miner. Eng.* **2016**, *96–97*, 83–93. [CrossRef]

55. Nagaraj, D.R.; Farinato, R. Chemical factor effects in saline and hypersaline waters in the flotation of Cu and Cu-Mo ores. In Proceedings of the XXVII International Mineral Processing Congress, Santiago, Chile, 20–24 October 2014; pp. 20–24.

56. Li, W.; Li, Y.; Wei, Z.; Xiao, Q.; Song, S. Fundamental Studies of SHMP in Reducing Negative Effects of Divalent Ions on Molybdenite Flotation. *Minerals* **2018**, *8*, 404. [CrossRef]

57. Rebolledo, E.; Laskowski, J.S.; Gutierrez, L.; Castro, S. Use of dispersants in flotation of molybdenite in seawater. *Miner. Eng.* **2017**, *100*, 71–74. [CrossRef]

58. Jeldres, R.I.; Arancibia-Bravo, M.P.; Reyes, A.; Aguirre, C.E.; Cortes, L.; Cisternas, L.A. The impact of seawater with calcium and magnesium removal for the flotation of copper-molybdenum sulphide ores. *Miner. Eng.* **2017**, *109*, 10–13. [CrossRef]

59. Chen, J.; Lan, L.; Liao, X. Depression effect of pseudo glycolythiourea acid in flotation separation of copper–molybdenum. *Trans. Nonferrous Met. Soc. China* **2013**, *23*, 824–831. [CrossRef]

60. Li, M.; Wei, D.; Shen, Y.; Liu, W.; Gao, S.; Liang, G. Selective depression effect in flotation separation of copper-molybdenum sulfides using 2,3-disulfanylbutanedioic acid. *Trans. Nonferrous Met. Soc. China* **2015**, *25*, 3126–3132. [CrossRef]

61. Yin, Z.; Sun, X.; Hu, Y.; Guan, Q.; Zhang, C.; Gao, Y.; Zhai, J. Depressing behaviors and mechanism of disodium bis (carboxymethyl) trithiocarbonate on separation of chalcopyrite and molybdenite. *Trans. Nonferrous Met. Soc. China* **2017**, *27*, 883–890. [CrossRef]

62. Li, M.; Wei, D.; Liu, Q.; Liu, W.; Zheng, J.; Sun, H. Flotation separation of copper-molybdenum sulfides using chitosan as a selective depressant. *Miner. Eng.* **2015**, *83*, 217–222. [CrossRef]

63. Yin, Z.; Sun, W.; Hu, Y.; Zhang, C.; Guan, Q.; Zhang, C. Separation of molybdenite from chalcopyrite in the presence of novel depressant 4-amino-3-thioxo-3,4-dihydro-1,2,4-triazin-5(2H)-one. *Minerals* **2017**, *7*, 146.

64. Yin, Z.; Sen, W.; Hu, Y.; Zhang, C.; Guan, Q.; Liu, R.; Chen, P.; Tian, M. Utilization of acetic acid-[(hydrazinylthioxomethyl)thio]-sodium as a novel selective depressant for chalcopyrite in the flotation separation of molybdenite. *Sep. Purif. Technol.* **2017**, *179*, 248–256. [CrossRef]

65. Huang, P.; Cao, M.; Liu, Q. Using chitosan as a selective depressant in the differential flotation Cu-Pb sulfides. *Int. J. Miner. Process.* **2012**, *106–109*, 8–15. [CrossRef]

66. Huang, P.; Cao, M.; Liu, Q. Adsorption of chitosan on chalcopyrite and galena from aqueous suspensions. *Colloids Surf. A Physicochem. Eng. Asp.* **2012**, *409*, 167–175. [CrossRef]

67. Crini, G.; Badot, P.-M. Application of chitosan, a natural aminopolysaccharide, for dye removal from aqueous solutions by adsorption processes using batch studies: A review of recent literature. *Prog. Polym. Sci.* **2008**, *33*, 399–447. [CrossRef]

68. Miller, J.D.; Ye, Y.; Jang, W.-H. Molybdenite Flotation from Copper Sulfide/Molybdenite Containing Materials by Ozone Conditioning. U.S. Patent No. 5,068,028, 26 November 1991.

69. Ye, Y.; Jang, W.H.; Yalamanchili, M.R.; Miller, J.D. Molybdenite flotation from copper/molybdenum concentrates by ozone conditioning. *Miner. Metall. Process.* **1990**, *7*, 173–179. [CrossRef]

70. Da Silva, L.M.; Jardim, W.F. Trends and strategies of ozone application in environmental problems. *Quim. Nova* **2006**, *29*, 310–317. [CrossRef]

71. Kasprzyk-Hordern, B.; Ziółek, M.; Nawrocki, J. Catalytic ozonation and methods of enhancing molecular ozone reactions in water treatment. *Appl. Catal. B Environ.* **2003**, *46*, 639–669. [CrossRef]

72. Huang, C.P.; Dong, C.; Tang, Z. Advanced chemical oxidation: Its present role and potential future in hazardous waste treatment. *Waste Manag.* **1993**, *13*, 361–377. [CrossRef]

73. Castro, S.; Lopez-Valdivieso, A.; Laskowski, J.S. Review of the flotation of molybdenite. Part I: Surface properties and floatability. *Int. J. Miner. Process.* **2016**, *148*, 48–58. [CrossRef]

74. Hirajima, T.; Mori, M.; Ichikawa, O.; Sasaki, K.; Miki, H.; Farahat, M.; Sawada, M. Selective flotation of chalcopyrite and molybdenite with plasma. *Miner. Eng.* **2014**, *66–68*, 102–111. [CrossRef]

75. Suyantara, G.P.W.; Hirajima, T.; Miki, H.; Sasaki, K.; Yamane, M.; Takida, E.; Kuroiwa, S.; Imaizumi, Y. Selective flotation of chalcopyrite and molybdenite using H_2O_2 oxidation method with the addition of ferrous sulfate. *Miner. Eng.* **2018**, *122*, 312–326. [CrossRef]

76. Lin, S.-S.; Gurol, M.D. Catalytic decomposition of hydrogen peroxide on iron oxide: Kinetics, mechanism, and implications. *Environ. Sci. Technol.* **1998**, *32*, 1417–1423. [CrossRef]

77. Li, X.; Huang, Y.; Li, C.; Shen, J.; Deng, Y. Degradation of pCNB by Fenton like process using α-FeOOH. *Chem. Eng. J.* **2015**, *260*, 28–36. [CrossRef]

78. Miki, H.; Matsuoka, H.; Hirajima, T.; Suyantara, G.P.W.; Sasaki, K. Electrolysis oxidation of chalcopyrite and molybdenite for selective flotation. *Mater. Trans.* **2017**, *58*, 761–767. [CrossRef]

79. Jha, R.K.T.; Satur, J.; Hiroyoshi, N.; Ito, M.; Tsunekawa, M. Carrier-microencapsulation using Si-catechol complex for suppressing pyrite floatability. *Miner. Eng.* **2008**, *21*, 889–893. [CrossRef]

80. Li, X.; Hiroyoshi, N.; Tabelin, C.B.; Naruwa, K.; Harada, C.; Ito, M. Suppressive effects of ferric-catecholate complexes on pyrite oxidation. *Chemosphere* **2019**, *214*, 70–78. [CrossRef]

81. Park, I.; Tabelin, C.B.; Magaribuchi, K.; Seno, K.; Ito, M.; Hiroyoshi, N. Suppression of the release of arsenic from arsenopyrite by carrier-microencapsulation using Ti-catechol complex. *J. Hazard. Mater.* **2018**, *344*, 322–332. [CrossRef]

82. Park, I.; Tabelin, C.B.; Seno, K.; Jeon, S.; Ito, M.; Hiroyoshi, N. Simultaneous suppression of acid mine drainage and arsenic release by carrier-microencapsulation using aluminum-catecholate complexes. *Chemosphere* **2018**, *205*, 414–425. [CrossRef] [PubMed]

83. Park, I.; Tabelin, C.B.; Jeon, S.; Li, X.; Seno, K.; Ito, M.; Hiroyoshi, N. A review of recent strategies for acid mine drainage prevention and mine tailings recycling. *Chemosphere* **2019**, *219*, 588–606. [CrossRef]

84. Park, I.; Tabelin, C.B.; Seno, K.; Jeon, S.; Inano, H.; Ito, M.; Hiroyoshi, N. Carrier-microencapsulation of arsenopyrite using Al-catecholate complex: Nature of oxidation products, effects on anodic and cathodic reactions, and coating stability under simulated weathering conditions. *Heliyon* **2020**, *6*, e03189. [CrossRef]

Article

Flotation Separation of Chalcopyrite and Molybdenite Assisted by Microencapsulation Using Ferrous and Phosphate Ions: Part I. Selective Coating Formation

Ilhwan Park [1],*, Seunggwan Hong [2], Sanghee Jeon [1], Mayumi Ito [1] and Naoki Hiroyoshi [1]

1   Division of Sustainable Resources Engineering, Faculty of Engineering, Hokkaido University, Sapporo 060-8628, Japan; shjun1121@eng.hokudai.ac.jp (S.J.); itomayu@eng.hokudai.ac.jp (M.I.); hiroyosi@eng.hokudai.ac.jp (N.H.)
2   Cooperative Program for Resources Engineering, Graduate School of Engineering, Hokkaido University, Sapporo 060-8628, Japan; seunggwan.hong.s4@elms.hokudai.ac.jp
*   Correspondence: i-park@eng.hokudai.ac.jp; Tel.: +81-11-706-6315

Received: 6 November 2020; Accepted: 9 December 2020; Published: 13 December 2020

Abstract: Porphyry Cu-Mo deposits, which are the most important sources of copper and molybdenum, are typically processed by flotation. In order to separate Cu and Mo minerals (mostly chalcopyrite and molybdenite), the strategy of depressing chalcopyrite while floating molybdenite has been widely adopted by using chalcopyrite depressants, such as NaHS, Na_2S, and Nokes reagent. However, these depressants are potentially toxic due to their possibility to emit H_2S gas. Thus, this study aims at developing a new concept for selectively depressing chalcopyrite via microencapsulation while using Fe^{2+} and $PO_4{}^{3-}$ forming $Fe^{(III)}PO_4$ coating. The cyclic voltammetry results indicated that Fe^{2+} can be oxidized to Fe^{3+} on the chalcopyrite surface, but not on the molybdenite surface, which arises from their different electrical properties. As a result of microencapsulation treatment using 1 mmol/L Fe^{2+} and 1 mmol/L $PO_4{}^{3-}$, chalcopyrite was much more coated with $FePO_4$ than molybdenite, which indicated that selective depression of chalcopyrite by the microencapsulation technique is highly achievable.

Keywords: porphyry deposits; chalcopyrite; molybdenite; flotation; microencapsulation; $FePO_4$ coating

1. Introduction

Porphyry Cu-Mo deposits are the most important sources of copper (Cu) and molybdenum (Mo), because they account for about 60% and 50% of the world's Cu and Mo productions [1]. In porphyry Cu-Mo deposits, chalcopyrite ($CuFeS_2$) and molybdenite (MoS_2) are the predominant Cu and Mo minerals [2,3], and they are separately recovered as Cu and Mo concentrates via a two-step process: (i) bulk flotation in order to produce Cu-Mo bulk concentrates; (ii) the selective flotation of molybdenite from Cu-Mo bulk concentrates. Bulk flotation is conducted with the assistance of proper collectors, such as xanthate for chalcopyrite and insoluble nonpolar oily collectors (e.g., diesel, kerosene, or fuel oil) for molybdenite [4,5], and pH adjuster (e.g., quick lime (CaO) or slaked lime ($Ca(OH)_2$) to increase the pulp pH at >10 where iron sulfides (mostly pyrite (FeS_2)) are depressed due to the competitive adsorption of OH^- that inhibits xanthate adsorption on the surface of iron sulfides [6]. After this, the Cu-Mo bulk concentrates are treated with chalcopyrite depressant (e.g., sodium hydrosulfide (NaHS), sodium sulfide (Na_2S), and Nokes reagent (P_2S_5 + NaOH)) in order to depress chalcopyrite while floating molybdenite [5]. These reagents function as chalcopyrite depressants by producing HS^- that desorbs the adsorbed xanthate on the chalcopyrite surface and/or reduces the pulp potential, in which chalcopyrite does not float [7].

Although the conventional chalcopyrite depressants are effective in separating Cu and Mo minerals from bulk Cu-Mo bulk concentrates, these have the potential to generate toxic hydrogen sulfide gas

($H_2S_{(g)}$) when pulp pH is not properly controlled [3,5,8]. At a pH below 10, for example, $H_2S_{(aq)}$ species starts forming and it is readily vaporized due to its high vapor pressure [7]. Thus, the flotation circuits should consist of covered flotation cells with an active ventilation system to avoid the accident that is caused by H_2S emission [5,7,9]. Moreover, the corrosive nature of H_2S that destroys pipelines and imperfect molybdenite recovery are other drawbacks of using conventional depressants [3,7,8,10].

Because of the above-mentioned limitations, there have been many attempts to replace the conventional chalcopyrite depressants with alternative depressants for molybdenite (e.g., dextrin [11,12], lignosulfonates [13], O-carboxymethyl chitosan [14,15], carboxymethylcellulose [16], humic acid [17], etc.) and chalcopyrite (e.g., sodium sulfite (Na_2SO_3) [3], pseudo-glycolythiourea acid (PGA) [18], 2,3-disulfanylbutanedioic acid (DMSA) [19], disodium bis (carboxymethyl) trithiocarbonate (DBT) [20], chitosan [21], etc.). Between these two approaches, the strategy of depressing chalcopyrite is more preferred than depressing molybdenite, because, in porphyry deposits, the amount of molybdenite is lower when compared to that of chalcopyrite (i.e., Cu grade, 0.44%; Mo grade, 0.018%) [2]. In the case that molybdenite depresses while chalcopyrite floats, the mechanical entrainment of molybdenite could occur within a large volume of froth products, resulting in the loss of valuable molybdenite, which is critical, because molybdenite recovery is of importance for Cu-Mo processing plants to be economically viable [22].

Miki et al. [9] reported that the difference in electrical resistivity between chalcopyrite and molybdenite can be utilized for reducing the floatability of chalcopyrite selectively. The electrical resistances of chalcopyrite and molybdenite are 234 Ω and 1.2–1.5 MΩ, respectively, so electrochemical reactions occur more preferably on the surface of chalcopyrite than that of molybdenite [9]. Based on these distinctively different electrical properties of minerals, this study investigated the application of microencapsulation technique while using ferrous (Fe^{2+}) and phosphate (PO_4^{3-}) ions as a pretreatment for depressing the floatability of chalcopyrite. Due to the extremely high electrical resistivity of molybdenite, Fe^{2+} is hardly oxidized to ferric ion (Fe^{3+}); however, chalcopyrite has a low electrical resistivity and, thus, Fe^{2+} can be oxidized to Fe^{3+}, which then reacts with PO_4^{3-} and forms ferric phosphate ($FePO_4$) on the surface of chalcopyrite, rendering it hydrophilic. This is the first part of a two-part paper, the aim of which is to investigate whether microencapsulation while using Fe^{2+} and PO_4^{3-} can selectively create $FePO_4$ coating on the surface of chalcopyrite rather than that of molybdenite. Specifically, the overall scope of this part is as follows: (i) elucidating the mechanism of selective coating formation via an electrochemical technique (i.e., cyclic voltammetry (CV)) and (ii) confirming whether chalcopyrite is selectively coated with $FePO_4$ via shake-flask experiments that are coupled with surface characterizations (i.e., scanning electron microscopy with energy-dispersive X-ray spectroscopy (SEM-EDX) and X-ray photoelectron spectroscopy (XPS)). A follow-up paper will discuss the effect of microencapsulation using Fe^{2+} and PO_4^{3-} on the selective depression of chalcopyrite in Cu-Mo flotation separation.

2. Materials and Methods

2.1. Mineral Samples

Chalcopyrite and molybdenite were obtained from Copper Queen Mine, Cochise County, AZ, USA and Spain Mine, Renfrew County, ON, Canada, respectively. The samples were crushed by a jaw crusher (BB 51, Retsch Inc. Haan, Germany), ground by a vibratory disc mill (RS 100, Retsch Inc., Haan, Germany), and then screened in order to obtain a size fraction of less than 75 μm. Mineralogical and chemical compositions of the samples were determined by X-ray diffraction (XRD; Figure 1) and X-ray fluorescence (XRF; Table 1). The chalcopyrite sample is composed of mainly chalcopyrite with pyrite and silicate minerals (e.g., quartz (SiO_2), amesite ($Mg_2Al_2SiO_5(OH)_4$), and actinolite ($Ca_2(Mg,Fe^{2+})_5Si_8O_{22}(OH)_2$)) as impurities, while molybdenite sample is highly pure and only contains trace amounts of impurities, as can be seen.

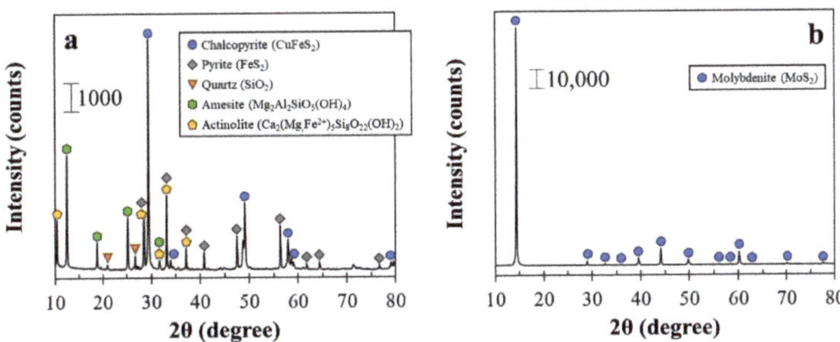

Figure 1. X-ray diffraction (XRD) patterns of (**a**) chalcopyrite and (**b**) molybdenite.

Table 1. Elemental compositions of chalcopyrite and molybdenite.

Chalcopyrite		Molybdenite	
Elements	wt.%	Elements	wt.%
Cu	23.2	Mo	56.8
Fe	32.6	S	43.0
S	29.4	Others	0.2
Si	9.5	-	-
Others	5.3	-	-

2.2. Stability of Fe^{2+} in the Presence of PO_4^{3-} vs. pH

The stability of Fe^{2+} in the presence of PO_4^{3-} was investigated as a function of pH in order to decide the suitable pH condition for $FePO_4$ coating formation on chalcopyrite to be selective. For this, a solution containing 1 mmol/L $FeSO_4 \cdot 7H_2O$ and 1 mmol/L KH_2PO_4 was prepared, and its pH was adjusted in the pH range of 2–6 while using dilute HCl and NaOH. All of the chemicals used in this study were of reagent grade (Wako Pure Chemical Industries, Ltd., Osaka, Japan). After pH adjustment, the solution was allowed to stabilize for 10 min. at room temperature under magnetic stirring (400 rpm). Afterwards, the solutions were filtered through 0.2 μm syringe-driven membrane filters (LMS Co. Ltd., Tokyo, Japan), and then analyzed by inductively coupled plasma atomic emission spectrometer (ICP-AES, ICPE-9820, Shimadzu Corporation, Kyoto, Japan) in order to identify the changes in dissolved Fe concentration.

2.3. Electrochemical Behaviors of Fe^{2+} on Chalcopyrite and Molybdenite

The electrochemical behaviors of Fe^{2+} on chalcopyrite and molybdenite were investigated by CV while using an electrochemical measurement unit (SI 1280B, Solartron Analytical, Farnborough, UK) with a conventional three-electrode system consisting of a mineral (chalcopyrite or molybdenite) working electrode, a platinum (Pt) counter electrode, and an Ag/AgCl (saturated KCl) reference electrode. The mineral electrodes were prepared in an identical way that was illustrated in our previous works [23,24]. Two types of electrolyte solutions were prepared: (i) 0.1 mol/L Na_2SO_4 and (ii) 1 mmol/L $FeSO_4 \cdot 7H_2O$ and 0.1 mol/L Na_2SO_4, both of which were adjusted to pH 4. The solution was equilibrated at 25 °C and then deoxygenated by N_2 purging for 30 min. After this, three electrodes were inserted and equilibrated at open circuit potential (OCP), and then CV was measured under the following conditions: initial scan polarity, positive; potential scan range, −0.8 to +0.8 V; scan rate, 30 mV/s.

2.4. Microencapsulation Treatment

Prior to the microencapsulation treatment, mineral samples were washed in order to remove the effects that are caused by the presence of oxidized layers formed during sample preparation.

The washing procedure is as follows: ultrasonic cleaning in ethanol, acid washing (1.0 M HNO_3), triple rinsing with de-ionized (DI) water (18.2 MΩ·cm), dewatering with acetone, and drying in a vacuum desiccator [25].

For the microencapsulation treatment, 1 g of washed mineral (e.g., chalcopyrite or molybdenite) and 10 mL of a solution containing 1 mmol/L Fe^{2+} and 1 mmol/L PO_4^{3-} (pH 4) were put into a 50-mL Erlenmeyer flask and then shaken in a constant temperature water bath at 25 °C and 120 min^{-1} for 1–6 h. After predetermined time intervals, the suspensions were filtered through 0.2 µm syringe-driven membrane filters and then analyzed by ICP-AES. Treated minerals were thoroughly washed with DI water, dried in a vacuum dry oven (40 °C), and then analyzed by SEM-EDX (JSM-IT200, JEOL Ltd., Tokyo, Japan) and XPS (JPS-9200, JEOL Ltd., Tokyo, Japan). The XPS analysis was conducted using an Al Kα X-ray source operated at 100 W (Voltage, 10 kV; Current, 10 mA) under ultrahigh vacuum conditions (approximately 10^{-7} Pa). The narrow scan spectra were calibrated while using the binding energy of adventitious carbon (C 1s) (285.0 eV) for charge correction. For deconvolutions of the spectra, XPSPEAK version 4.1 (Raymond WM Kwok, Chinese University of Hong Kong, Hong Kong, China) was used with an 80% Gaussian–20% Lorentzian peak model and a true Shirley background [26,27].

3. Results

3.1. Stability of Fe^{2+} in the Presence of PO_4^{3-} vs. pH

The oxidation rate of Fe^{2+} to Fe^{3+} by dissolved oxygen (DO) is known to be pH-dependent [28,29]. Fe^{2+} oxidation rate under acidic conditions (i.e., pH < 4) is very slow and independent of pH, as shown in Figure 2a. However, at pH $\geq$ 5 Fe^{2+}, the Fe^{2+} oxidation rate shows the second order dependence on [OH^-], indicating that a 100-fold increase in the rate occurs for a unit increase in pH [28]. In the studied system, not only Fe^{2+}, but also PO_4^{3-} coexist, so the stability of Fe^{2+} in the presence of PO_4^{3-} as a function of pH was investigated (Figure 2b). The concentration of Fe^{2+} was almost not changed between pH 2 and 4, but above which Fe^{2+} concentration decreases rapidly. This result is identical to that reported by previous works [28,29], indicating that Fe^{2+} is stable at pH < 4, even in the presence of PO_4^{3-}. The oxidation of Fe^{2+} by DO is not preferred, because the selective coating formation could only be achieved when Fe^{2+} oxidation occurs on the surface of chalcopyrite. Thus, pH 4 was selected for microencapsulation treatment to be selective.

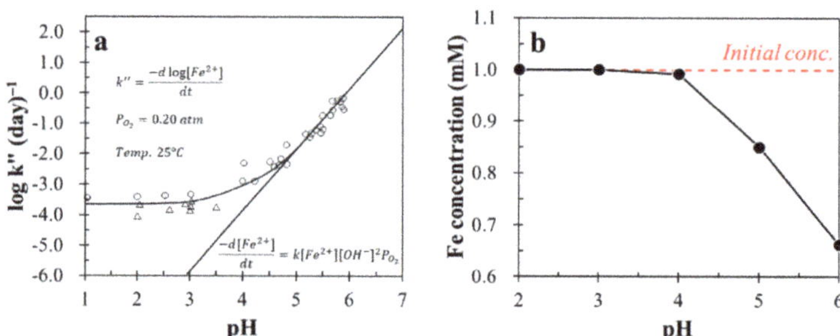

Figure 2. (**a**) Oxidation rate of ferrous iron species as a function of pH (reprinted with permission from Morgan and Lahav [29], copyright (2007) Elsevier) and (**b**) the change in dissolved Fe concentration ([Fe^{2+}]$_{initial}$ = 1 mmol/L) in the presence of 1 mmol/L PO_4^{3-} as a function of pH.

3.2. Electrochemical Behavior of Fe^{2+} on Chalcopyrite and Molybdenite

Figure 3 shows the cyclic voltammograms of chalcopyrite (Figure 3a) and molybdenite (Figure 3b) in the absence and presence of 1 mmol/L Fe^{2+}. The current density was continuously increased as the applied potential increased, which means that chalcopyrite as well as its oxidation products (e.g., Fe^{2+}) were oxidized (Equations (1) and (2)), as shown in the first anodic scan of chalcopyrite in the absence of Fe^{2+} (Figure 3(a2)) [30–32]. When 1 mmol/L Fe^{2+} was present, the current density was apparently increased as compared to that without Fe^{2+}. This increase in current density most likely resulted from the oxidation of Fe^{2+} to Fe^{3+} (Equation (2)).

$$CuFeS_2 \rightarrow Cu_{(1-x)}Fe_{(1-y)}S_{(2-z)} + xCu^{2+} + yFe^{2+} + zS^0 + 2(x+y)e^- \tag{1}$$

$$Fe^{2+} \rightarrow Fe^{3+} + e^- \tag{2}$$

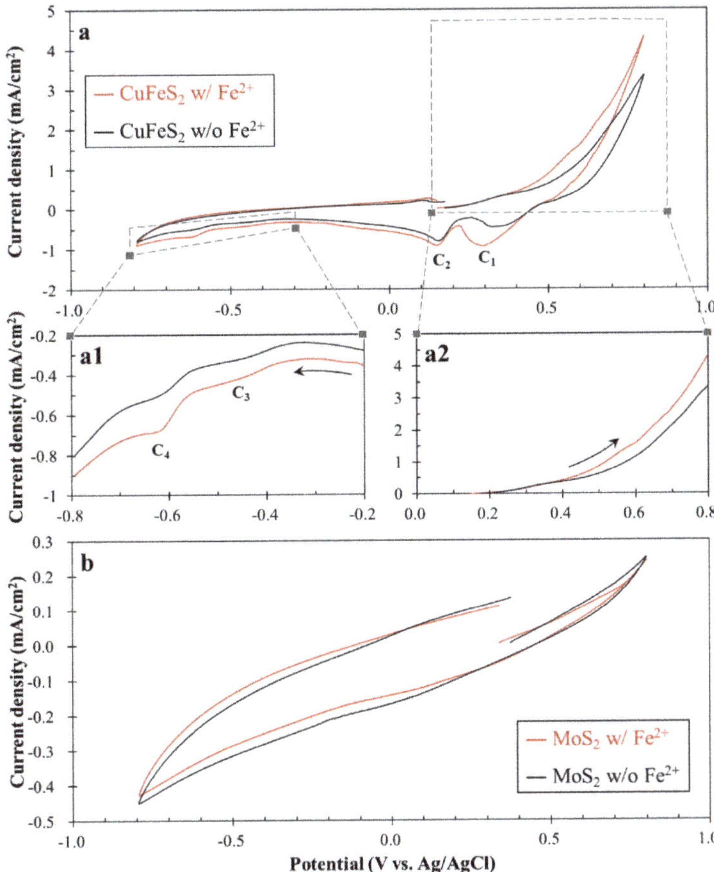

Figure 3. Cyclic voltammograms of (**a**) chalcopyrite (**a1**: cathodic scan from −0.2 to −0.8 V, **a2**: anodic scan from open circuit potential (OCP) to 0.8 V) and (**b**) molybdenite in the absence and presence of 1 mmol/L Fe^{2+}. Note that the arrows in Figure 3(a1,a2) denote the sweep direction.

During the cathodic scan in the absence of 1 mmol/L Fe^{2+}, it exhibited four cathodic peaks (C_1, C_2, C_3, and C_4), as shown in Figure 3a,(a1). According to Holiday and Richmond [31], the first two cathodic peaks (C_1 and C_2) were due to the reduction of dissolved species, like Fe^{3+} and Cu^{2+}. Holiday and Richmond [31] executed cathodic linear-sweep voltammetry while using stationary and rotating chalcopyrite electrodes, both of which were pretreated at 0.65 V vs. SCE for 2 min., and found out that two cathodic peaks at 0.38 and 0.15 V vs. SCE observed in the voltammogram using a stationary electrode were absent when the electrode was rotated. Thus, the appearance of C_1 and C_2 can be explained by the reduction of Fe^{3+} to Fe^{2+} at 0.3–0.4 V (Equation (3)) and the formation of covellite

(CuS) at 0.1–0.2 V (Equation (4)), respectively. At the applied potential between −0.4 and −0.7 V, two additional peaks (C_3 and C_4) were observed (Figure 3(a1)), which resulted from the reduction of CuS to chalcocite (Cu_2S) and S^0 to H_2S, as illustrated in Equations (5) and (6) [30–32].

$$Fe^{3+} + e^- \rightarrow Fe^{2+} \tag{3}$$

$$Cu^{2+} + S^0 + 2e^- \rightarrow CuS \tag{4}$$

$$2CuS + 2H^+ + 2e^- \rightarrow Cu_2S + H_2S \tag{5}$$

$$S^0 + 2H^+ + 2e^- \rightarrow H_2S \tag{6}$$

Similarly, these four cathodic peaks (C_1, C_2, C_3, and C_4) were also observed in the voltammogram in the presence of 1 mmol/L Fe^{2+}; however, it is important to note that the current density of C_1 was obviously increased. This indicates that more Fe^{3+} was generated during the anodic scan of chalcopyrite in the presence of 1 mmol/L Fe^{2+} as compared to that of control. On the other hand, cyclic voltammograms of molybdenite showed that there is no clear difference between the absence and presence of 1 mmol/L Fe^{2+} (Figure 3b). It is most likely attributed to the low electrical conductivity of molybdenite, making the electrochemical reactions of Fe^{2+}/Fe^{3+} redox couple hard to occur on its surface. Therefore, the cyclic voltammetry results imply that the selective oxidation of Fe^{2+} on the chalcopyrite surface is highly possible.

3.3. Microencapsulation Treatment for Chalcopyrite and Molybdenite

Figure 4 shows the precipitation rates of dissolved Fe and P during microencapsulation treatment for chalcopyrite and molybdenite. The precipitation rate is calculated by the following equation:

$$\text{Precipitation rate (\%)} = (Ci - Ct)/Ci \times 100\% \tag{7}$$

where C_i is the initial concentration of dissolved Fe or P in mg/L, and C_t is the concentration of dissolved Fe or P in mg/L at time t. As shown in Figure 4a, the precipitation of dissolved Fe considerably occurred in the presence of chalcopyrite; that is, around 60% of Fe^{2+} was precipitated after 1 h treatment and it reached 93% after 6 h. Similarly, dissolved P was also significantly precipitated with chalcopyrite (Figure 4b). When compared to the precipitation rate of dissolved P, the precipitated amount of dissolved Fe was a bit low. This is probably due to chalcopyrite dissolution that releases $Fe^{2+/3+}$, resulting in the lowering of the Fe precipitation rate. These results indicate that Fe^{2+} is oxidized to Fe^{3+} on the surface of chalcopyrite, then, the resultant product (i.e., Fe^{3+}) reacts with $H_2PO_4^-$, a dominant species of phosphate at pH 4, forming $FePO_4$, as shown in the following equation:

$$Fe^{3+} + H_2PO_4^- \rightarrow FePO_4\downarrow + 2H^+ \tag{8}$$

It is interesting to note that, even during microencapsulation treatment for molybdenite, around 20–40% of dissolved Fe and P were precipitated (Figure 4a,b). Although the CV results indicated that Fe^{2+} oxidation could not occur on the surface of molybdenite (Figure 3b), the precipitation of dissolved Fe and P apparently occurred in the presence of molybdenite. These opposite results are attributed to the different electrical resistivity of different sides (e.g., basal and edge planes) of molybdenite [9]. Miki et al. [9] investigated electrolysis oxidation treatment while using basal- and edge-plane-oriented molybdenite electrodes and confirmed that electron transfer was actively pursued through the edge plane when compared to that through the basal plane. At an applied potential of 1.0 V vs. SHE, for example, the current density of the edge-plane electrode was around 0.1 mA/cm^2, lasting for 800 s, while it was closed to 0 mA/cm^2 in the case of the basal-plane electrode. The particle size of molybdenite used in this study is less than 75 µm. As the size of molybdenite particle reduces, the ratio of basal-plane/edge-plane also decreases [33]. This means that electron transfer reactions, like Fe^{2+} oxidation on fine molybdenite, become easier to occur when compared to coarse molybdenite. Thus, approximately 20–40% of dissolved Fe and P were precipitated during the reaction with molybdenite (Figure 4).

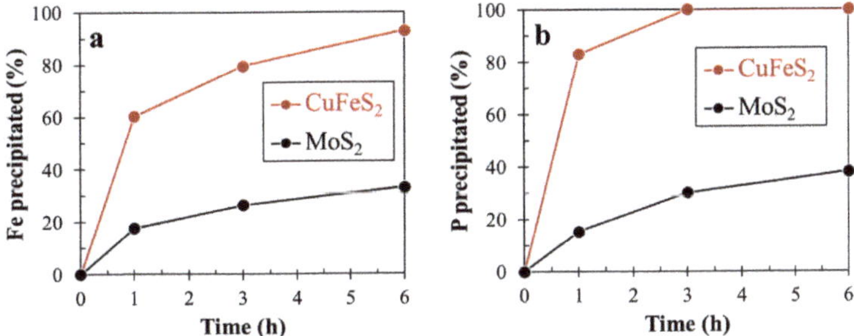

Figure 4. Precipitation rates of dissolved Fe (**a**) and P (**b**).

Figure 5 shows the SEM-EDX results of untreated- and treated-chalcopyrite. As can be seen in the SEM image of untreated chalcopyrite, its surface was smooth and clear; however, after microencapsulation treatment, the surface morphology dramatically changed the secondary precipitates that covered the surface of chalcopyrite. Untreated chalcopyrite exhibited strong signals of Cu, Fe, and S, while treated chalcopyrite showed that the signals of Cu and S were decreased, but signals of Fe and O were increased, as shown in the EDX spectra of these samples. In addition, the P signal appeared in the spectrum of treated chalcopyrite. These increased (i.e., Fe, P, and O) and decreased (i.e., Cu and S) signals were also found in the elemental maps of treated chalcopyrite. These results indicate that, after microencapsulation treatment, Fe–P–O-containing coatings were formed on the chalcopyrite surface. On the other hand, the SEM-EDX results of molybdenite with and without treatment (Figure 6) showed that there is no clear difference between the two samples, although 40% of dissolved Fe and P were precipitated (Figure 4). It could be speculated that the signals of Fe and P are almost noise levels due to the small amount of precipitate present on the molybdenite surface or the precipitates do not exist on the molybdenite surface.

In order to further characterize the surfaces of untreated and treated minerals, XPS analysis was adopted, which can analyze a very thin layer of coating (around ~6 nm) and give the information on the chemical state of the element. Figure 7a shows the XPS P 2p spectra of untreated and treated chalcopyrite. As it can be seen, treated chalcopyrite exhibited a broad peak that was centered at around 133.5 eV composed of adsorbed PO_4^{3-} (130.0, 131.4 and 132.8 eV) [34] and $P^{(V)}$ of $FePO_4$ (133.7 eV) [34,35], which support that $FePO_4$ coating was formed on the chalcopyrite surface by microencapsulation while using Fe^{2+} and PO_4^{3-}. In the case of the XPS spectrum of treated molybdenite (Figure 7b), a weak and gentle peak of $FePO_4$ was observed. The peak area of $FePO_4$ in the XPS spectrum of treated chalcopyrite was three times higher than that of treated molybdenite, which indicated that more $FePO_4$ is present on the chalcopyrite surface, as confirmed by Figures 4–6. All of the results that were obtained in this study imply that Fe^{2+} oxidation, followed by $FePO_4$ precipitation, preferably occurs on the chalcopyrite surface rather than on the molybdenite surface, so the selective depression of chalcopyrite in the flotation of bulk concentrates is highly achievable, which will be evaluated in detail in Part 2 of this study.

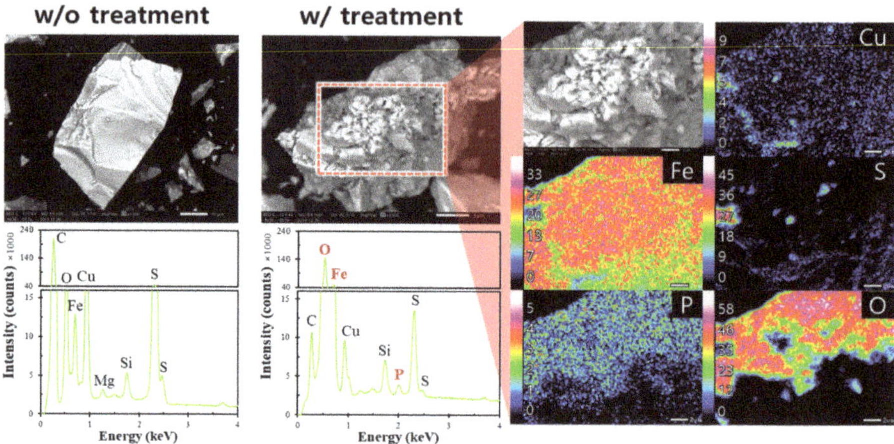

Figure 5. Scanning electron microscopy with energy-dispersive X-ray spectroscopy (SEM-EDX) results of chalcopyrite with and without microencapsulation treatment.

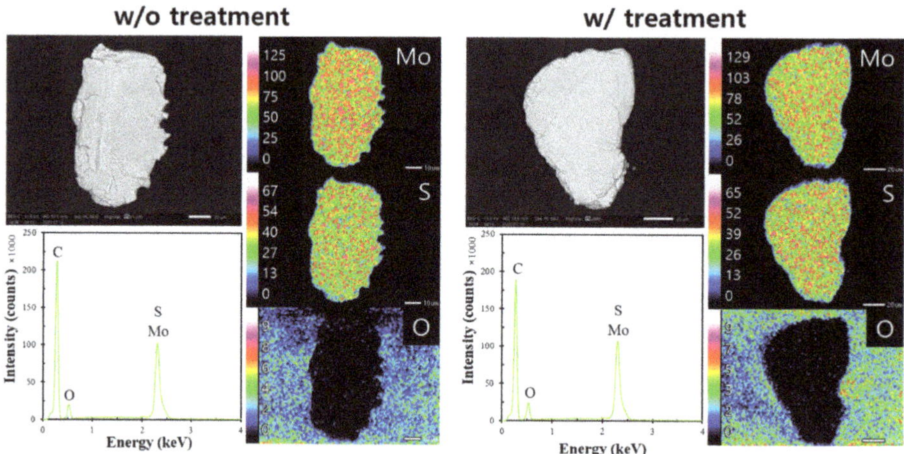

Figure 6. SEM-EDX results of molybdenite with and without microencapsulation treatment.

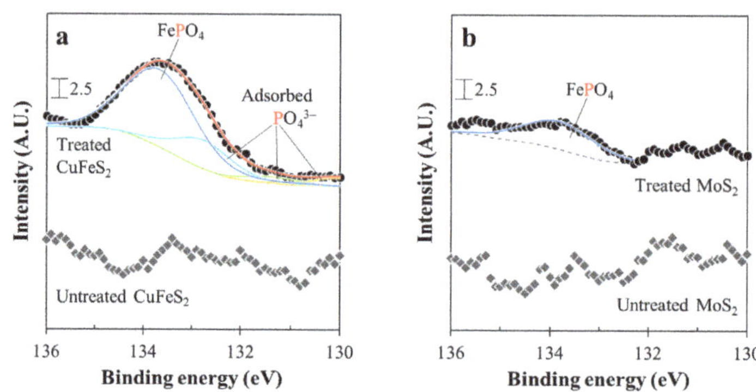

Figure 7. X-ray photoelectron spectroscopy (XPS) P 2p spectra of (**a**) chalcopyrite and (**b**) molybdenite with and without microencapsulation treatment.

4. Conclusions

This study investigated microencapsulation while using Fe^{2+} and PO_4^{3-} in order to selectively coat chalcopyrite with $FePO_4$ with the aim of improving Cu-Mo flotation separation. The findings of this study are summarized, as follows:

1. In the presence of phosphate ion, Fe^{2+} was stable at pH $\leq$ 4, above which, however, Fe^{2+} became unstable due to its rapid oxidation to Fe^{3+}, which was then precipitated as $FePO_4$ and/or $FeO(OH)$. Thus, microencapsulation treatment using Fe^{2+} and PO_4^{3-} is recommended to be conducted at pH 4 in order to achieve the selective Fe^{2+} oxidation on chalcopyrite surface.

2. The CV results indicated that Fe^{2+} oxidation can occur on the chalcopyrite surface, but not on the molybdenite surface, due to their different electrical properties.

3. After microencapsulation treatment using Fe^{2+} and PO_4^{3-}, SEM-EDX and XPS analyses confirmed that chalcopyrite was more coated with $FePO_4$ than molybdenite.

Author Contributions: Conceptualization, I.P., S.J., M.I. and N.H.; methodology, I.P.; investigation, I.P. and S.H.; data curation, I.P. and S.H.; writing—original draft preparation, I.P.; writing—review and editing, I.P., S.H., S.J., M.I. and N.H. All authors have read and agreed to the published version of the manuscript.

Funding: This study is financially supported by Japan Oil, Gas and Metals National Corporation (JOGMEC).

Conflicts of Interest: The authors declare no conflict of interest.

References

1. Bulatovic, S.M. 12-Flotation of Copper Sulfide Ores. In *Handbook of Flotation Reagents*; Bulatovic, S.M., Ed.; Elsevier: Amsterdam, The Netherlands, 2007; pp. 235–293. ISBN 978-0-444-53029-5.

2. Ayuso, R.A.; Barton, M.D.; Blakely, R.J.; Bodnar, R.J.; Dilles, J.H.; Gray, F.; Graybeal, F.T.; Mars, J.L.; McPhee, D.K.; Seal II, R.R.; et al. *Porphyry Copper Deposit Model: Chapter B in Mineral Deposit Models for Resource Assessment*; Scientific Investigations Report; U.S. 2010-5070-B; Geological Survey: Reston, VA, USA, 2010.

3. Miki, H.; Hirajima, T.; Muta, Y.; Suyantara, G.P.W.; Sasaki, K. Effect of Sodium Sulfite on Floatability of Chalcopyrite and Molybdenite. *Minerals* **2018**, *8*, 172. [CrossRef]

4. Johnston, A.; Meadows, D.G.; Cappuccitti, F. Copper Mineral Processing. In *SME Mineral Processing & Extractive Metallurgy Handbook*; Society for Mining, Metallurgy & Exploration (SME): Englewood, CO, USA, 2019; Volume 2, pp. 1615–1641.

5. Amelunxen, P.; Schmitz, C.; Hill, H.; Goodweiler, N.; Andres, J. Molybdenum. In *SME Mineral Processing & Extractive Metallurgy Handbook*; Society for Mining, Metallurgy & Exploration (SME): Englewood, CO, USA, 2019; Volume 2, pp. 1891–1916.

6. Sutherland, K.L.; Wark, I.W. Depressants. In *Principles of Flotation*; Australasian Institute of Mining and Metallurgy: Melbourne, Australia, 1955; pp. 113–153.

7. Park, I.; Hong, S.; Jeon, S.; Ito, M.; Hiroyoshi, N. A Review of Recent Advances in Depression Techniques for Flotation Separation of Cu–Mo Sulfides in Porphyry Copper Deposits. *Metals* **2020**, *10*, 1269. [CrossRef]

8. Hirajima, T.; Miki, H.; Suyantara, G.P.W.; Matsuoka, H.; Elmahdy, A.M.; Sasaki, K.; Imaizumi, Y.; Kuroiwa, S. Selective flotation of chalcopyrite and molybdenite with H_2O_2 oxidation. *Miner. Eng.* **2017**, *100*, 83–92. [CrossRef]

9. Miki, H.; Matsuoka, H.; Hirajima, T.; Suyantara, G.P.W.; Sasaki, K. Electrolysis Oxidation of Chalcopyrite and Molybdenite for Selective Flotation. *Mater. Trans.* **2017**, *58*, 761–767. [CrossRef]

10. Yin, Z.; Sun, W.; Hu, Y.; Zhang, C.; Guan, Q.; Zhang, C. Separation of Molybdenite from Chalcopyrite in the Presence of Novel Depressant 4-Amino-3-thioxo-3,4-dihydro-1,2,4-triazin-5(2H)-one. *Minerals* **2017**, *7*, 146. [CrossRef]

11. Wie, J.M.; Fuerstenau, D.W. The effect of dextrin on surface properties and the flotation of molybdenite. *Int. J. Miner. Process.* **1974**, *1*, 17–32. [CrossRef]

12. Jorjani, E.; Barkhordari, H.R.; Tayebi Khorami, M.; Fazeli, A. Effects of aluminosilicate minerals on copper–molybdenum flotation from Sarcheshmeh porphyry ores. *Miner. Eng.* **2011**, *24*, 754–759. [CrossRef]

13. Ansari, A.; Pawlik, M. Floatability of chalcopyrite and molybdenite in the presence of lignosulfonates. Part II. Hallimond tube flotation. *Miner. Eng.* **2007**, *20*, 609–616. [CrossRef]

14. Yuan, D.; Cadien, K.; Liu, Q.; Zeng, H. Flotation separation of Cu-Mo sulfides by O-Carboxymethyl chitosan. *Miner. Eng.* **2019**, *134*, 202–205. [CrossRef]

15. Yuan, D.; Cadien, K.; Liu, Q.; Zeng, H. Adsorption characteristics and mechanisms of O-Carboxymethyl chitosan on chalcopyrite and molybdenite. *J. Colloid Interface Sci.* **2019**, *552*, 659–670. [CrossRef]

16. Kor, M.; Korczyk, P.M.; Addai-Mensah, J.; Krasowska, M.; Beattie, D.A. Carboxymethylcellulose Adsorption on Molybdenite: The Effect of Electrolyte Composition on Adsorption, Bubble–Surface Collisions, and Flotation. *Langmuir* **2014**, *30*, 11975–11984. [CrossRef] [PubMed]

17. Yuan, D.; Cadien, K.; Liu, Q.; Zeng, H. Selective separation of copper-molybdenum sulfides using humic acids. *Miner. Eng.* **2019**, *133*, 43–46. [CrossRef]

18. Chen, J.; Lan, L.; Liao, X. Depression effect of pseudo glycolythiourea acid in flotation separation of copper–molybdenum. *Trans. Nonferrous Met. Soc. China* **2013**, *23*, 824–831. [CrossRef]

19. Li, M.; Wei, D.; Shen, Y.; Liu, W.; Gao, S.; Liang, G. Selective depression effect in flotation separation of copper–molybdenum sulfides using 2,3-disulfanylbutanedioic acid. *Trans. Nonferrous Met. Soc. China* **2015**, *25*, 3126–3132. [CrossRef]

20. Yin, Z.; Sun, W.; Hu, Y.; Guan, Q.; Zhang, C.; Gao, Y.; Zhai, J. Depressing behaviors and mechanism of disodium bis (carboxymethyl) trithiocarbonate on separation of chalcopyrite and molybdenite. *Trans. Nonferrous Met. Soc. China* **2017**, *27*, 883–890. [CrossRef]

21. Li, M.; Wei, D.; Liu, Q.; Liu, W.; Zheng, J.; Sun, H. Flotation separation of copper–molybdenum sulfides using chitosan as a selective depressant. *Miner. Eng.* **2015**, *83*, 217–222. [CrossRef]

22. Laskowski, J.S.; Castro, S.; Ramos, O. Effect of seawater main components on frothability in the flotation of Cu-Mo sulfide ore. *Physicochem. Probl. Miner. Process.* **2013**, 17–29. [CrossRef]

23. Park, I.; Tabelin, C.B.; Seno, K.; Jeon, S.; Inano, H.; Ito, M.; Hiroyoshi, N. Carrier-microencapsulation of arsenopyrite using Al-catecholate complex: Nature of oxidation products, effects on anodic and cathodic reactions, and coating stability under simulated weathering conditions. *Heliyon* **2020**, *6*, e03189. [CrossRef]

24. Park, I.; Tabelin, C.B.; Seno, K.; Jeon, S.; Ito, M.; Hiroyoshi, N. Simultaneous suppression of acid mine drainage formation and arsenic release by Carrier-microencapsulation using aluminum-catecholate complexes. *Chemosphere* **2018**, *205*, 414–425. [CrossRef]

25. McKibben, M.A.; Barnes, H.L. Oxidation of pyrite in low temperature acidic solutions: Rate laws and surface textures. *Geochim. Cosmochim. Acta* **1986**, *50*, 1509–1520. [CrossRef]

26. Shirley, D.A. High-Resolution X-Ray Photoemission Spectrum of the Valence Bands of Gold. *Phys. Rev. B* **1972**, *5*, 4709–4714. [CrossRef]

27. Nesbitt, H.W.; Muir, I.J. X-ray photoelectron spectroscopic study of a pristine pyrite surface reacted with water vapour and air. *Geochim. Cosmochim. Acta* **1994**, *58*, 4667–4679. [CrossRef]

28. Stumm, W.; Morgan, J.J. *Aquatic Chemistry: Chemical Equilibria and Rates in Natural Waters*; Wiley-Interscience: New York, NY, USA, 1996.

29. Morgan, B.; Lahav, O. The effect of pH on the kinetics of spontaneous Fe(II) oxidation by O_2 in aqueous solution–basic principles and a simple heuristic description. *Chemosphere* **2007**, *68*, 2080–2084. [CrossRef] [PubMed]

30. Qin, W.Q.; Yang, C.R.; Wang, J.; Zhang, Y.; Jiao, F.; Zhao, H.B.; Zhu, S. Effect of Fe^{2+} and Cu^{2+} Ions on the Electrochemical Behavior of Massive Chalcopyrite in Bioleaching System. *Adv. Mater. Res.* **2013**, *825*, 472–476. [CrossRef]

31. Holliday, R.I.; Richmond, W.R. An electrochemical study of the oxidation of chalcopyrite in acidic solution. *J. Electroanal. Chem. Interfacial Electrochem.* **1990**, *288*, 83–98. [CrossRef]

32. Liang, C.-L.; Xia, J.-L.; Yang, Y.; Nie, Z.-Y.; Zhao, X.-J.; Zheng, L.; Ma, C.-Y.; Zhao, Y.-D. Characterization of the thermo-reduction process of chalcopyrite at 65 °C by cyclic voltammetry and XANES spectroscopy. *Hydrometallurgy* **2011**, *107*, 13–21. [CrossRef]

33. Castro, S.; Lopez-Valdivieso, A.; Laskowski, J.S. Review of the flotation of molybdenite. Part I: Surface properties and floatability. *Int. J. Miner. Process.* **2016**, *148*, 48–58. [CrossRef]

34. Park, I.; Higuchi, K.; Tabelin, C.B.; Jeon, S.; Ito, M.; Hiroyoshi, N. Suppression of arsenopyrite oxidation by microencapsulation using ferric-catecholate complexes and phosphate. *Chemosphere*, under review.
35. Zeng, L.; Li, X.; Shi, Y.; Qi, Y.; Huang, D.; Tadé, M.; Wang, S.; Liu, S. $FePO_4$ based single chamber air-cathode microbial fuel cell for online monitoring levofloxacin. *Biosens. Bioelectron.* **2017**, *91*, 367–373. [CrossRef]

Publisher's Note: MDPI stays neutral with regard to jurisdictional claims in published maps and institutional affiliations.

Article

Flotation Separation of Chalcopyrite and Molybdenite Assisted by Microencapsulation Using Ferrous and Phosphate Ions: Part II. Flotation

Ilhwan Park [1,*], Seunggwan Hong [2], Sanghee Jeon [1], Mayumi Ito [1] and Naoki Hiroyoshi [1]

1 Division of Sustainable Resources Engineering, Faculty of Engineering, Hokkaido University, Sapporo 060-8628, Japan; shjun1121@eng.hokudai.ac.jp (S.J.); itomayu@eng.hokudai.ac.jp (M.I.); hiroyosi@eng.hokudai.ac.jp (N.H.)
2 Cooperative Program for Resources Engineering, Graduate School of Engineering, Hokkaido University, Sapporo 060-8628, Japan; seunggwan.hong.s4@elms.hokudai.ac.jp
* Correspondence: i-park@eng.hokudai.ac.jp; Tel.: +81-11-706-6315

Abstract: Porphyry-type deposits are the major sources of copper and molybdenum, and flotation has been adopted to recover them separately. The conventional reagents used for depressing copper minerals, such as NaHS, Na_2S, and Nokes reagent, have the potential to emit toxic H_2S gas when pulp pH was not properly controlled. Thus, in this study the applicability of microencapsulation (ME) using ferrous and phosphate ions as an alternative process to depress the floatability of chalcopyrite was investigated. During ME treatment, the use of high concentrations of ferrous and phosphate ions together with air introduction increased the amount of $FePO_4$ coating formed on the chalcopyrite surface, which was proportional to the degree of depression of its floatability. Although ME treatment also reduced the floatability of molybdenite, ~92% Mo could be recovered by utilizing emulsified kerosene. Flotation of chalcopyrite/molybdenite mixture confirmed that the separation efficiency was greatly improved from 10.9% to 66.8% by employing ME treatment as a conditioning process for Cu-Mo flotation separation.

Keywords: porphyry deposits; chalcopyrite; molybdenite; microencapsulation; flotation

Citation: Park, I.; Hong, S.; Jeon, S.; Ito, M.; Hiroyoshi, N. Flotation Separation of Chalcopyrite and Molybdenite Assisted by Microencapsulation Using Ferrous and Phosphate Ions: Part II. Flotation. *Metals* **2021**, *11*, 439. https://doi.org/10.3390/met11030439

Academic Editor: Man Seung Lee

Received: 19 February 2021
Accepted: 5 March 2021
Published: 7 March 2021

Publisher's Note: MDPI stays neutral with regard to jurisdictional claims in published maps and institutional affiliations.

1. Introduction

Porphyry-type deposits are the major sources of copper (Cu) and molybdenum (Mo)—approximately 60% of Cu and 50% of Mo are annually produced from these deposits [1–3]. Apart from Cu and Mo, precious metals (e.g., gold (Au), silver (Ag), and platinum group elements (PGEs)) and several strategic/high-tech elements (e.g., rhenium (Re), tungsten (W), bismuth (Bi), indium (In), tellurium (Te), and selenium (Se)) may reach economic concentrations, thus being recovered as by-products during porphyry ore processing [3,4]. Typically, porphyry-type deposits are developed via a series of processes: (i) open-pit mining to excavate the ores, (ii) closed-circuit comminution to liberate valuable and non-valuable minerals, (iii) bulk flotation to recover Cu and Mo minerals (i.e., mainly chalcopyrite ($CuFeS_2$) and molybdenite (MoS_2)) as Cu-Mo bulk concentrates, and (iv) Mo flotation to separate Cu and Mo minerals from bulk concentrates [2].

At the final stage of porphyry ore processing (i.e., flotation separation of Cu and Mo minerals), Cu-Mo bulk concentrates are conditioned with Cu depressants (e.g., sodium hydrosulfide (NaHS), sodium sulfide (Na_2S), and Nokes reagent (P_2S_5 + NaOH)) to reduce the floatability of chalcopyrite while floating molybdenite [1,2]. Although effective, these reagents have the potential to emit hydrogen sulfide (H_2S) gas—known as toxic and deadly gas—when the pulp pH is not properly maintained; for example, at pH < 10, HS^- ion derived from Cu depressants starts forming $H_2S_{(aq)}$ species, which is then readily transformed into the gaseous phase, i.e., $H_2S_{(g)}$, due to its extremely high vapor pressure (P_{H_2S} = 20.03 atm at 25 °C) [2,5]. To avoid the accident that is caused by H_2S emission, the flotation circuits should consist of covered flotation cells together with active ventilation systems as well as a NaOH-solution scrubber treating any process off-gas [6]. However, the use of covered flotation cells obviously obstructs visual inspection of the froth, which lowers the efficiency of operations of flotation process [6]. Not only this, but the use of above-mentioned reagents may destroy the pipelines due to the corrosive nature of H_2S and yield imperfect molybdenite recovery [7–9], so the attention should be paid to the development of alternative techniques.

This paper is Part II of a two-part basic study to investigate how the application of microencapsulation technique affects flotation behaviors of chalcopyrite and molybdenite. Microencapsulation is a technique that encloses the target material with coatings composed of small discrete solid particles or small liquid droplets. It has been reported that microencapsulation using redox-sensitive compounds has an ability to selectively form the coating on the surface of conductive minerals [10–14]. In Part I of this study, microencapsulation using ferrous and phosphate ions was investigated with the aim of creating ferric phosphate ($FePO_4$) coating selectively on the chalcopyrite surface [15]. Electrochemical study revealed that ferrous oxidation occurred preferably on the surface of chalcopyrite rather than molybdenite. Moreover, the results of shake-flask experiments coupled with surface characterizations were consistent with electrochemical study; that is, ferrous oxidation followed by $FePO_4$ formation occurred predominantly on the chalcopyrite surface. However, it remains unclear how $FePO_4$ coating formed via microencapsulation affects the floatability of chalcopyrite and molybdenite. In Part II, thus, flotation tests of chalcopyrite and/or molybdenite with and without microencapsulation treatment were carried out to evaluate its effect on the floatability of chalcopyrite and molybdenite as well as their separation efficiency.

2. Materials and Methods

2.1. Mineral Samples

Chalcopyrite and molybdenite were obtained from Copper Queen Mine, Cochise County, AZ, USA and Spain Mine, Renfrew County, ON, Canada, respectively. The samples were crushed by a jaw crusher (BB 51, Retsch Inc. Haan, Germany), ground by a vibratory disc mill (RS 100, Retsch Inc., Haan, Germany), and then screened in order to obtain a size fraction of 38–75 μm. Chalcopyrite sample mainly consists of chalcopyrite (~70%) with minor amounts of impurities like pyrite (FeS_2) and silicate minerals (e.g., quartz (SiO_2), amesite ($Mg_2Al_2SiO_5(OH)_4$), and actinolite ($Ca_2(Mg, Fe^{2+})_5Si_8O_{22}(OH)_2)$)), while molybdenite sample is highly pure (~99.8%). The detailed sample characterizations can be found in Part I of this study [15].

2.2. Microencapsulation Treatment

Prior to microencapsulation (ME) treatment, mineral samples were deslimed by ultrasonication in ethanol for 1 min, followed by decantation after 1 min sedimentation. Afterward, the sediments were rinsed with acetone four times and dried in vacuum desiccators.

Microencapsulation treatments were conducted using an agitator-type flotation machine (FT-1000, Heiko, Japan). In a 400-mL flotation cell, 20 g of mineral sample (i.e., chalcopyrite, molybdenite, or chalcopyrite/molybdenite mixture (1:1, w/w)) and 200 mL of coating solution containing ferrous and phosphate ions were mixed at 1000 rpm for 1 h. Coating solution was prepared using $FeSO_4 \cdot 7H_2O$ and KH_2PO_4, and its pH was adjusted to 4.0 ± 0.1 while using dilute HCl and NaOH. All the chemicals used in this study were of reagent grade (Wako Pure Chemical Industries, Ltd., Osaka, Japan). To investigate the suitable conditions for depressing the floatability of chalcopyrite, the effects of the concentrations of ferrous and phosphate ions (1 and 10 mM) and the introduction of air (1 L/min) during ME treatment were examined.

After ME treatment, the suspension was filtered and washed with deionized (DI) water 4 times to remove the remaining ferrous and phosphate ions, and then used for flotation experiments. Filtrates were collected using 0.2 μm syringe-driven filters (LMS Co. Ltd.,

Tokyo, Japan) and analyzed by inductively coupled plasma atomic emission spectrometer (ICP-AES, ICPE9820, Shimadzu Corporation, Kyoto, Japan) in order to identify the changes in dissolved Fe and P concentrations.

2.3. Flotation Tests

Flotation tests were conducted using an agitator-type flotation machine (FT-1000, Heiko, Japan) equipped with a 400-mL flotation cell in which 20 g of washed sample, 10 g of quartz (99.9% SiO_2, Wako Pure Chemical Industries, Ltd., Osaka, Japan), and 400 mL DI water were added. The purpose of adding quartz is to measure the amounts of chalcopyrite/molybdenite recovered by entrainment. The pulp was suspended at 1000 rpm for 3 min, and then conditioned with 25 μL/L of frother (methyl isobutyl carbinol, MIBC, Tokyo Chemical Industry Co., Ltd., Tokyo, Japan) for another 3 min. In the flotation of molybdenite and chalcopyrite/molybdenite mixture, 2.5 L/t of emulsified kerosene was added as a collector for molybdenite and conditioned for 3 min. Emulsified kerosene was prepared as follows: (i) kerosene (Wako Pure Chemical Industries, Ltd., Osaka, Japan) was mixed with distilled water in the concentration of 20 wt.%; (ii) emulsification was carried out using an ultrasonic homogenizer (ULTRA-TURRAX, IKA, Königswinter, Germany) for 60 s [16,17]. Afterward, air was injected into the suspension at a flow rate of 1 L/min, and flotation was carried out for 3 min (for the case of flotation of chalcopyrite/molybdenite mixture, it was conducted for up to 6 min). Froth products and tailings obtained after flotation were weighed after drying at 105 °C for 24 h, and their elemental compositions were determined using X-ray fluorescence spectrometer (XRF, EDXL300, Rigaku Corporation, Tokyo, Japan).

2.4. Contact Angle Measurements

Contact angle measurements were carried out in order to estimate the changes in the surface wettability of molybdenite before and after ME treatment (with and without kerosene addition). For this, molybdenite specimen was cut using a diamond cutter to obtain a small cuboid crystal (~5 mm (w) × 5 mm (d) × 10 mm (h)), which was then polished using a polishing machine (SAPHIR 250 M1, ATM GmbH, Mammelzen, Germany) with a series of silicon carbide papers (P320, P600, and P1200) and diamond suspensions (3 and 1 μm). Afterward, the contact angles of (i) untreated molybdenite, (ii) ME-treated molybdenite with 10 mM Fe^{2+}/$H_2PO_4^-$ at 1 L/min air introduction for 1 h, and (iii) ME-treated molybdenite with conditioning using 2.5 L/t of emulsified kerosene for 3 min were measured by a high-magnification digital microscope (VHX-1000, Keyence Corporation, Osaka, Japan) with built in image analysis capability. Each measurement was repeated 3 times at different spots on the mineral surface to ascertain that the differences observed were statistically significant.

3. Results and Discussion

3.1. Flotation of Chalcopyrite

Figure 1 shows the effect of ME treatment on the floatability of chalcopyrite. For the case of untreated chalcopyrite, about 77% of Cu was recovered after 3 min flotation even in the absence of any collector. When chalcopyrite was treated by ME with 1 mM Fe^{2+} and 1 mM $H_2PO_4^-$ prior to flotation, Cu recovery was almost the same as the one without ME treatment. In Part I of this study [15], it was confirmed that after ME treatment using 1 mM Fe^{2+} and 1 mM $H_2PO_4^-$, chalcopyrite was obviously coated with $FePO_4$; however, its floatability was not affected by $FePO_4$ coating.

One of the possible reasons for why ME treatment did not affect the floatability of chalcopyrite could be due to the insufficient amount of $FePO_4$ coating for depressing chalcopyrite. After 1 h ME treatment, ~80% of dissolved Fe and P were precipitated as $FePO_4$, indicating that the wt.% of $FePO_4$ present on the chalcopyrite surface was approximately 0.12% [15]. If chalcopyrite is covered with a large amount of $FePO_4$ coating (i.e., >0.12%), its floatability may be decreased. Thus, an attempt was made to increase the amount of

FePO$_4$ coating formed on the chalcopyrite surface by increasing the concentrations of Fe^{2+} and H$_2$PO$_4^-$ from 1 to 10 mM during ME treatment. As shown in Figure 2, Cu recovery was decreased from ~80% to ~70% when Fe^{2+}/H$_2$PO$_4^-$ concentrations increased from 1 to 10 mM. Although increasing Fe^{2+}/H$_2$PO$_4^-$ concentrations during ME treatment could reduce the floatability of chalcopyrite, the degree of chalcopyrite depression is not enough for separating chalcopyrite and molybdenite.

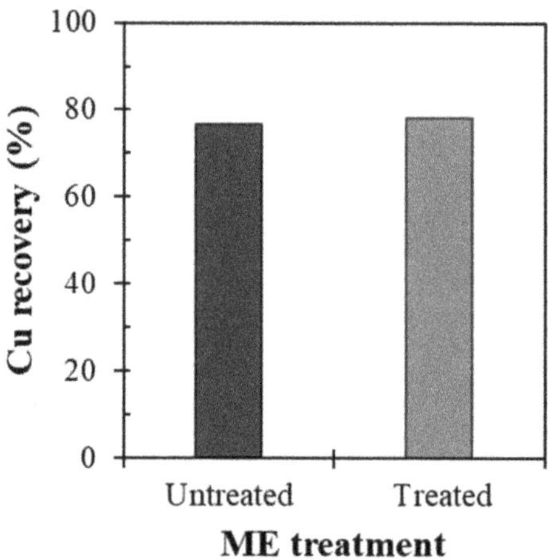

Figure 1. Effect of ME treatment on the floatability of chalcopyrite.

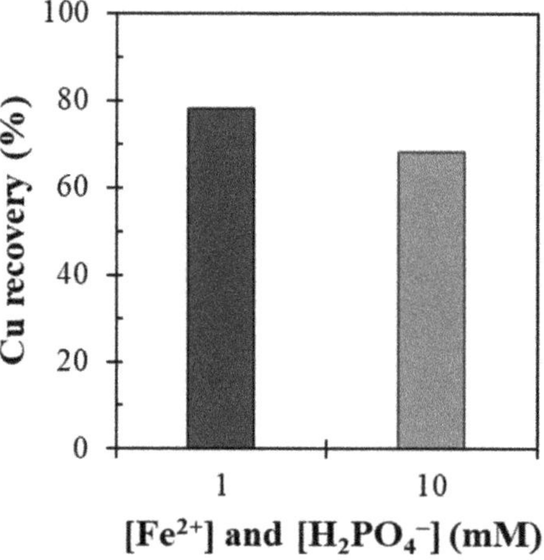

Figure 2. Effect of the concentration of Fe^{2+} and H$_2$PO$_4^-$ on the floatability of chalcopyrite.

The concentration of FePO$_4$ formed after ME treatment was 0.66 mM for the case using 1 mM Fe^{2+}/H$_2$PO$_4^-$, while ~1 mM of FePO$_4$ was formed when 10 mM Fe^{2+}/H$_2$PO$_4^-$ was used. These results support our deduction that increasing the amount of FePO$_4$ coating can enhance the depression of chalcopyrite. However, there are large amounts of unreacted ferrous and phosphate ions that remained in the solution after ME treatment. As illustrated in Equation (1), oxygen is an essential reactant for ferrous oxidation.

$$2Fe^{2+} + 2H^+ + 1/2O_2 = 2Fe^{3+} + H_2O \tag{1}$$

To facilitate the formation of more coatings on the surface of chalcopyrite, thus, the air was introduced during ME treatment in the presence of 10 mM $Fe^{2+}/H_2PO_4^-$. As shown in Figure 3, air introduction during ME treatment obviously improved the depression of chalcopyrite floatability; that is, Cu recovery decreased from about 70% (without air introduction) to <15% (with air introduction). Control experiment—flotation of chalcopyrite treated under air introduction (1 L/min) for 1 h in the absence of $Fe^{2+}/H_2PO_4^-$—confirmed that air introduction did not play an important role in depressing the floatability of chalcopyrite because Cu recovery was almost the same as that of untreated chalcopyrite (data not shown). The main role of introducing air is to promote the cathodic half-cell reaction occurring on the surface of chalcopyrite (i.e., oxygen reduction reaction; Equation (2)), thereby enhancing the anodic half-cell reaction (i.e., ferrous oxidation reaction; Equation (3)). When the air was injected during ME treatment, ~4 mM of $FePO_4$ was formed, indicating that the formation of large amounts of $FePO_4$ coating is effective in depressing chalcopyrite.

$$2H^+ + 1/2O_2 + 2e^- = H_2O \tag{2}$$

$$Fe^{2+} = Fe^{3+} + e^- \tag{3}$$

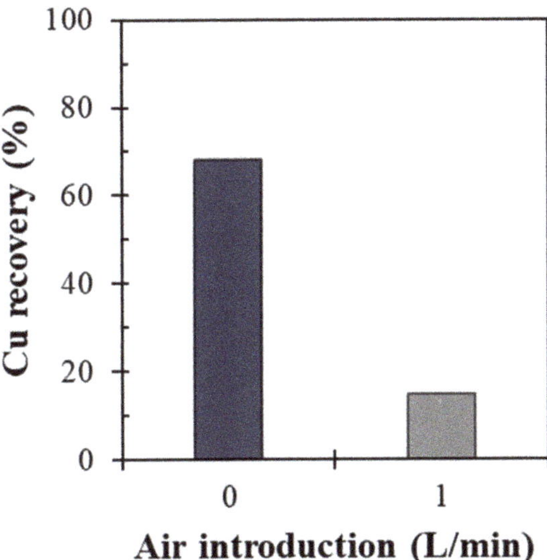

Figure 3. Effect of air introduction during ME treatment with 10 mM $Fe^{2+}/H_2PO_4^-$ on the floatability of chalcopyrite.

3.2. Flotation of Molybdenite

The main purpose of ME treatment is to depress the floatability of chalcopyrite, so molybdenite was also treated by ME under the same conditions where chalcopyrite was effectively depressed (i.e., 10 mM $Fe^{2+}/H_2PO_4^-$; 1 L/min air introduction). As shown in Figure 4, ME treatment had a detrimental effect on the floatability of molybdenite; that is, after ME treatment, Mo recovery decreased from ~45% to ~6%. For the flotation separation of chalcopyrite and molybdenite, the former should be depressed while floating the latter; thus, this is an unwelcome result because molybdenite was also depressed after ME treatment. As confirmed by XPS analysis of ME-treated chalcopyrite and molybdenite shown in Part I of this study [15], the amount of $FePO_4$ coating formed on the molybdenite surface was obviously smaller than that formed on the chalcopyrite surface; however, molybdenite was strongly depressed as much as chalcopyrite was (Figures 3 and 4).

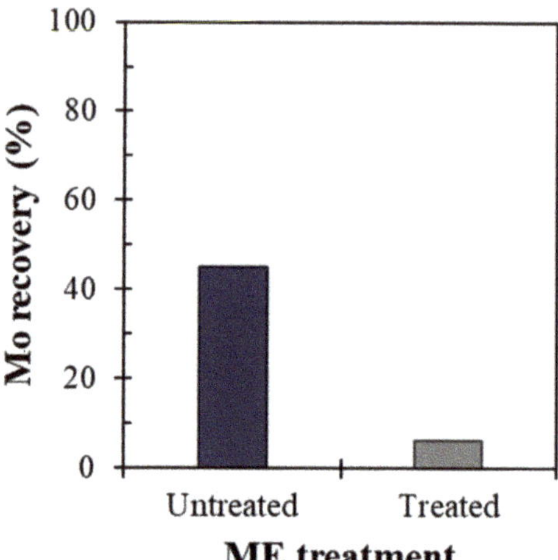

Figure 4. Effect of ME treatment on the floatability of molybdenite. Note that ME treatment was conducted with 10 mM $Fe^{2+}/H_2PO_4^-$ and 1 L/min air introduction.

The presence of precipitates on the molybdenite surface has been reported to reduce its floatability; for example, (i) flotation of Cu-Mo ores in seawater where seawater precipitates (e.g., $Mg(OH)_2$ and $CaCO_3$) formed at pH > 9.5 are accumulated on the molybdenite surface [18–21]; (ii) flotation of chalcopyrite/molybdenite mixture after plasma pretreatment resulting in the formation of Cu/Fe oxyhydroxides on the surface of molybdenite [22]. Hirajima and coworkers [20–22] reported that the reduced floatability of molybdenite caused by precipitates can be healed by adding kerosene—a commonly used collector for molybdenite. In this study, thus, kerosene was also adopted after ME treatment to improve the recovery of molybdenite. As can be seen in Figure 5, the addition of kerosene was remarkably effective in improving the floatability of ME-treated molybdenite; that is, Mo recovery was increased from ~6% to ~92% when 2.5 L/t of kerosene was added.

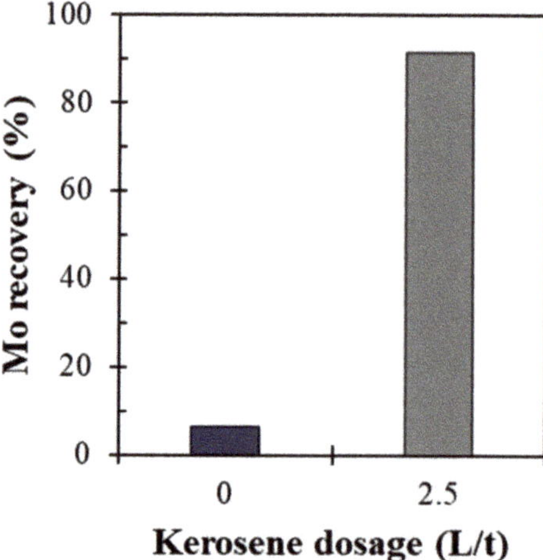

Figure 5. Effect of kerosene dosage on the floatability of ME-treated molybdenite.

The increased Mo recovery achieved by kerosene addition most likely resulted from the improvement of hydrophobicity of molybdenite surface. As illustrated in Figure 6, the contact angle of untreated molybdenite was ~94° but decreased to ~39° after ME treatment due to the presence of hydrophilic $FePO_4$ precipitates on its surface. When ME-treated molybdenite reacted with kerosene, however, the contact angle increased to ~109°, indicating that the hydrophobicity of molybdenite was improved, even better than that of bare molybdenite surface. This increased hydrophobicity of ME-treated molybdenite can be explained as follows: kerosene is adsorbed on molybdenite surface where $FePO_4$ is not present and/or covers not only molybdenite but also the attached $FePO_4$ precipitates.

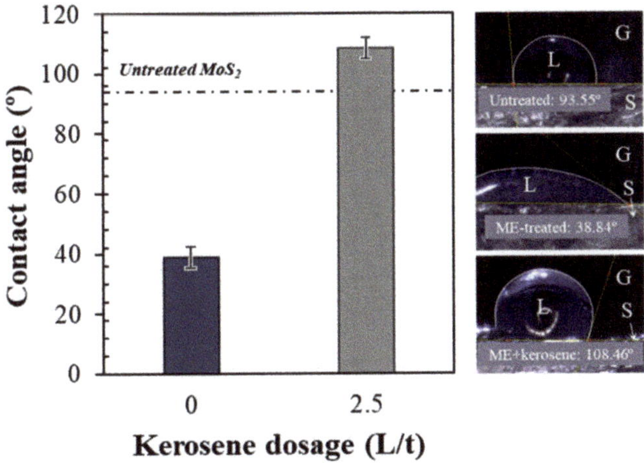

Figure 6. Effect of kerosene dosage on the contact angle of ME-treated molybdenite. Note that G, L, and S on the right side of the photos indicate gas (air), liquid (water droplet), and solid (molybdenite) phases, respectively.

Based on the results of Sections 3.1 and 3.2, Cu-Mo flotation separation could be achievable under the following conditions: ME treatment with 10 mM $Fe^{2+}/H_2PO_4^-$ while introducing air at 1 L/min air introduction, and flotation with 2.5 L/t kerosene. Although ME treatment followed by flotation of single minerals (i.e., chalcopyrite or molybdenite) looked promising, flotation results can be differed when they are mixed, so the following section deals with the mixed minerals system.

3.3. Flotation of Chalcopyrite/Molybdenite Mixture

Figure 7 shows flotation results of chalcopyrite/molybdenite mixture with and without ME treatment. As illustrated in Figure 7a, flotation of untreated chalcopyrite/molybdenite mixture showed that both minerals were floated well and, after 6 min flotation, the recovery of Cu and Mo reached about 83% and 92%, respectively. On the other hand, Cu and Mo recoveries after 6 min flotation of ME-treated mixture were ~33% and ~93%, respectively (Figure 7b). In the case of ME-treated mixture, froth products with higher Mo grade (46.33–49.3%) and lower Cu grade (3.8–5.2%) compared to those of untreated mixture (i.e., Mo grade, 33.8–34.6%; Cu grade, 9.6–9.8%) were obtained. These results indicate that the application of ME treatment prior to flotation of chalcopyrite/molybdenite mixture could selectively depress the floatability of chalcopyrite.

The effect of ME treatment on flotation of chalcopyrite/molybdenite mixture was evaluated by the classical first-order flotation kinetic model (Equation (4)) [23]:

$$R(t) = R_\infty \left[1 - \exp(-kt)\right], \tag{4}$$

where $R(t)$ and R_∞ are the recovery of chalcopyrite/molybdenite at time t and an infinite time, and k is the first-order rate constant. A nonlinear least square regression was used to calculate R_∞ and k from the best fit with experimental data. The obtained R_∞ and k were used for calculating the modified rate constant (Equation (5)) and the selectivity index of mineral I over mineral II (Equation (6)) [24]:

$$K_M = R_\infty \cdot k,\tag{5}$$

$$S.I.\ (I/II) = (K_M\ of\ mineral\ I)/(K_M\ of\ mineral\ II).\tag{6}$$

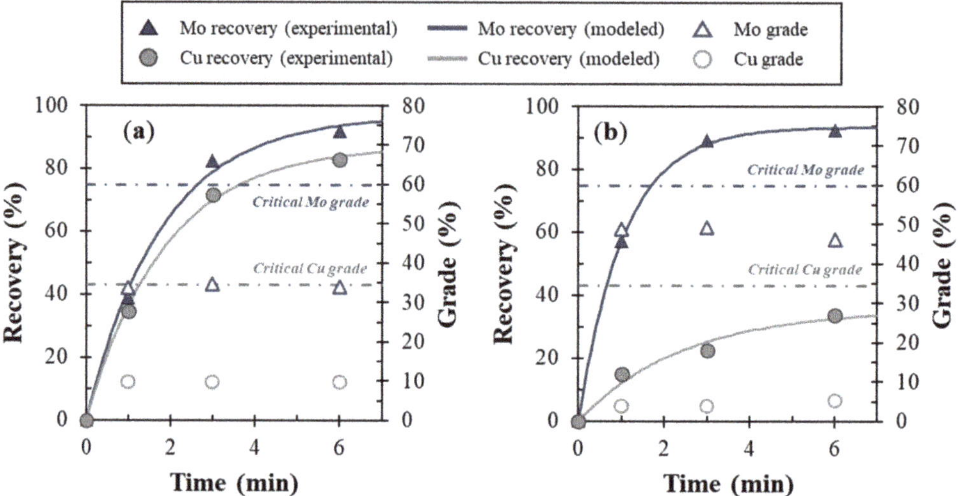

Figure 7. Effect of ME treatment on the floatability of chalcopyrite and molybdenite in the mixed mineral flotation: (**a**) untreated and (**b**) treated chalcopyrite/molybdenite mixture. Note that makers indicate experimental data while lines denote calculated values based on the first-order flotation kinetic model (Equation (4)).

The regression coefficients (e.g., R_∞ and k), K_M, and S.I. (Mo/Cu) were summarized in Table 1. The maximum recovery (R_∞) of molybdenite was almost the same irrespective of ME treatment (i.e., 97.0% without ME; 93.5% with ME), whereas R_∞ of ME-treated chalcopyrite decreased significantly from 87.3% to 35.7%. After ME treatment, the rate constant (k) and modified rate constant (K_M) increased for molybdenite but decreased for chalcopyrite. Moreover, the selectivity index of Mo/Cu of ME-treated chalcopyrite/molybdenite mixture was about five-fold higher than that of untreated mixture, indicating that ME treatment has an ability to selectively depress the floatability of chalcopyrite.

Table 1. Non-linear regression results for the first-order kinetic model fitting to the experimental data (Figure 6).

Parameters	Untreated		Treated	
	Chalcopyrite	Molybdenite	Chalcopyrite	Molybdenite
R^2	0.99	0.99	0.93	0.99
R_∞ (%)	87.3	97.0	35.7	93.5
k (min^{-1})	0.54	0.56	0.41	0.96
K_M (min^{-1})	0.47	0.54	0.15	0.89
S.I. (Mo/Cu)	1.16		6.08	

Figure 8 shows the relationship between Mo recovery in the froth and Cu recovery in the tailing after flotation with and without ME treatment, which is often used for the assessment of the separation efficiency (Equation (7)) [16,25–27]:

$$\eta = R_c - (1 - R_t),\tag{7}$$

where η is Newton's efficiency, R_c is the recovery of molybdenite in the froth, and R_t is the recovery of chalcopyrite in the tailing. For the case of untreated chalcopyrite/molybdenite mixture, the separation efficiencies were in the range of 4.0–10.9% (Figure 8). Compared to this, the application of ME treatment greatly improved the separation efficiency. Until 3 min of flotation, Mo recovery was rapidly increased up to ~90% but, afterward, slightly increased up to 93.5%. On the other hand, Cu recovery was apparently lower than Mo recovery but continuously increased up to 35.7% with time, which results in the highest separation efficiency at 3 min. In short, the separation efficiencies of untreated and treated mixture obtained after 3 min flotation were 10.9% and 66.8%, respectively. This suggests that the application of ME treatment as a Cu depression process prior to Cu-Mo flotation separation is effective in improving Mo/Cu separation efficiency.

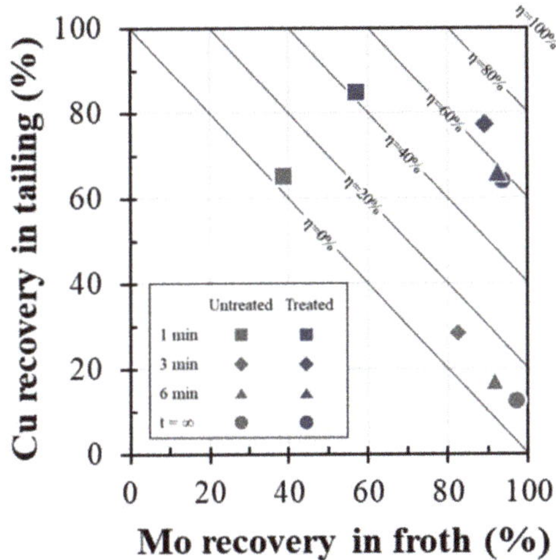

Figure 8. Relationship between Mo recovery in froth and Co recovery in tailing obtained in the mixed mineral flotation.

4. Conclusions

This study investigated the effect of microencapsulation using $Fe^{2+}/H_2PO_4^-$ as a pretreatment for Cu-Mo flotation separation. The findings of this study are summarized, as follows:

1. ME treatment using 10 mM $Fe^{2+}/H_2PO_4^-$ had a negligible effect on the depression of chalcopyrite floatability, but air introduction during ME treatment dramatically reduced Cu recovery from ~70% to ~15%. The air introduction played an important role in enhancing ferrous oxidation occurring on the surface of chalcopyrite, thereby improving the formation of $FePO_4$ coating on its surface.
2. Not only chalcopyrite, but the floatability of molybdenite was also depressed after ME treatment. The reduced floatability of ME-treated molybdenite, however, could be improved by utilizing emulsified kerosene during flotation.
3. The application of ME treatment was also effective for mixed minerals system that the separation efficiency increased from 10.9% (without ME treatment) to 66.8% (with ME treatment).

Author Contributions: Conceptualization, I.P., S.J., M.I. and N.H.; methodology, I.P.; investigation, I.P. and S.H.; data curation, I.P. and S.H.; writing—original draft preparation, I.P.; writing—review and editing, I.P., S.H., S.J., M.I. and N.H.; funding acquisition, I.P. All authors have read and agreed to the published version of the manuscript.

Funding: This study is financially supported by Japan Oil, Gas and Metals National Corporation (JOGMEC).

Institutional Review Board Statement: Not applicable.

Informed Consent Statement: Not applicable.

Data Availability Statement: Data available on request due to restrictions, as the research is ongoing.

Conflicts of Interest: The authors declare no conflict of interest.

References

1. Bulatovic, S.M. 12-Flotation of Copper Sulfide Ores. In *Handbook of Flotation Reagents*; Bulatovic, S.M., Ed.; Elsevier: Amsterdam, The Netherlands, 2007; pp. 235–293. ISBN 978-0-444-53029-5.
2. Park, I.; Hong, S.; Jeon, S.; Ito, M.; Hiroyoshi, N. A Review of Recent Advances in Depression Techniques for Flotation Separation of Cu–Mo Sulfides in Porphyry Copper Deposits. *Metals* **2020**, *10*, 1269. [CrossRef]
3. John, D.A.; Ayuso, R.A.; Barton, M.D.; Blakely, R.J.; Bodnar, R.J.; Dilles, J.H.; Gray, F.; Graybeal, F.T.; Mars, J.L.; McPhee, D.K.; et al. *Porphyry Copper Deposit Model: Chapter B in Mineral Deposit Models for Resource Assessment*; Scientific Investigations Report 2010-5070-B; U.S. Geological Survey: Menlo Park, CA, USA, 2010; 169p.
4. Cioacă, M.-E.; Munteanu, M.; Lynch, E.P.; Arvanitidis, N.; Bergqvist, M.; Costin, G.; Ivanov, D.; Milu, V.; Arvidsson, R.; Iorga-Pavel, A.; et al. Mineralogical Setting of Precious Metals at the Assarel Porphyry Copper-Gold Deposit, Bulgaria, as Supporting Information for the Development of New Drill Core 3D XCT-XRF Scanning Technology. *Minerals* **2020**, *10*, 946. [CrossRef]
5. Stull, D.R. Vapor Pressure of Pure Substances. Organic and Inorganic Compounds. *Ind. Eng. Chem.* **1947**, *39*, 517–540. [CrossRef]
6. Amelunxen, P.; Schmitz, C.; Hill, H.; Goodweiler, N.; Andres, J. Molybdenum. In *SME Mineral Processing & Extractive Metallurgy Handbook*; Society for Mining, Metallurgy & Exploration (SME): Englewood, CO, USA, 2019; Volume 2, pp. 1891–1916.
7. Hirajima, T.; Miki, H.; Suyantara, G.P.W.; Matsuoka, H.; Elmahdy, A.M.; Sasaki, K.; Imaizumi, Y.; Kuroiwa, S. Selective Flotation of Chalcopyrite and Molybdenite with H_2O_2 Oxidation. *Miner. Eng.* **2017**, *100*, 83–92. [CrossRef]
8. Miki, H.; Matsuoka, H.; Hirajima, T.; Suyantara, G.P.W.; Sasaki, K. Electrolysis Oxidation of Chalcopyrite and Molybdenite for Selective Flotation. *Mater. Trans.* **2017**, *58*, 761–767. [CrossRef]
9. Yin, Z.; Sun, W.; Hu, Y.; Zhang, C.; Guan, Q.; Zhang, C. Separation of Molybdenite from Chalcopyrite in the Presence of Novel Depressant 4-Amino-3-Thioxo-3,4-Dihydro-1,2,4-Triazin-5(2H)-One. *Minerals* **2017**, *7*, 146. [CrossRef]
10. Park, I.; Tabelin, C.B.; Magaribuchi, K.; Seno, K.; Ito, M.; Hiroyoshi, N. Suppression of the Release of Arsenic from Arsenopyrite by Carrier-Microencapsulation Using Ti-Catechol Complex. *J. Hazard. Mater.* **2018**, *344*, 322–332. [CrossRef] [PubMed]
11. Park, I.; Tabelin, C.B.; Seno, K.; Jeon, S.; Ito, M.; Hiroyoshi, N. Simultaneous Suppression of Acid Mine Drainage Formation and Arsenic Release by Carrier-Microencapsulation Using Aluminum-Catecholate Complexes. *Chemosphere* **2018**, *205*, 414–425. [CrossRef] [PubMed]
12. Li, X.; Hiroyoshi, N.; Tabelin, C.B.; Naruwa, K.; Harada, C.; Ito, M. Suppressive Effects of Ferric-Catecholate Complexes on Pyrite Oxidation. *Chemosphere* **2019**, *214*, 70–78. [CrossRef] [PubMed]
13. Park, I.; Tabelin, C.B.; Seno, K.; Jeon, S.; Inano, H.; Ito, M.; Hiroyoshi, N. Carrier-Microencapsulation of Arsenopyrite Using Al-Catecholate Complex: Nature of Oxidation Products, Effects on Anodic and Cathodic Reactions, and Coating Stability under Simulated Weathering Conditions. *Heliyon* **2020**, *6*, e03189. [CrossRef] [PubMed]
14. Park, I.; Higuchi, K.; Tabelin, C.B.; Jeon, S.; Ito, M.; Hiroyoshi, N. Suppression of Arsenopyrite Oxidation by Microencapsulation Using Ferric-Catecholate Complexes and Phosphate. *Chemosphere* **2021**, *269*, 129413. [CrossRef] [PubMed]
15. Park, I.; Hong, S.; Jeon, S.; Ito, M.; Hiroyoshi, N. Flotation Separation of Chalcopyrite and Molybdenite Assisted by Microencapsulation Using Ferrous and Phosphate Ions: Part I. Selective Coating Formation. *Metals* **2020**, *10*, 1667. [CrossRef]
16. Hornn, V.; Ito, M.; Shimada, H.; Tabelin, C.B.; Jeon, S.; Park, I.; Hiroyoshi, N. Agglomeration-Flotation of Finely Ground Chalcopyrite and Quartz: Effects of Agitation Strength during Agglomeration Using Emulsified Oil on Chalcopyrite. *Minerals* **2020**, *10*, 380. [CrossRef]
17. Hornn, V.; Ito, M.; Shimada, H.; Tabelin, C.B.; Jeon, S.; Park, I.; Hiroyoshi, N. Agglomeration–Flotation of Finely Ground Chalcopyrite Using Emulsified Oil Stabilized by Emulsifiers: Implications for Porphyry Copper Ore Flotation. *Metals* **2020**, *10*, 912. [CrossRef]
18. Castro, S. Challenges in flotation of Cu–Co sulfide ores in sea water. In *The First International Symposium on Water in Mineral Processing*; Society for Mining, Metallurgy & Exploration (SME): Littleton, CO, USA, 2012; pp. 29–40.
19. Jeldres, R.I.; Arancibia-Bravo, M.P.; Reyes, A.; Aguirre, C.E.; Cortes, L.; Cisternas, L.A. The Impact of Seawater with Calcium and Magnesium Removal for the Flotation of Copper-Molybdenum Sulphide Ores. *Miner. Eng.* **2017**, *109*, 10–13. [CrossRef]

20. Hirajima, T.; Suyantara, G.P.W.; Ichikawa, O.; Elmahdy, A.M.; Miki, H.; Sasaki, K. Effect of Mg^{2+} and Ca^{2+} as Divalent Seawater Cations on the Floatability of Molybdenite and Chalcopyrite. *Miner. Eng.* **2016**, *96–97*, 83–93. [CrossRef]
21. Suyantara, G.P.W.; Hirajima, T.; Miki, H.; Sasaki, K. Floatability of Molybdenite and Chalcopyrite in Artificial Seawater. *Miner. Eng.* **2018**, *115*, 117–130. [CrossRef]
22. Hirajima, T.; Mori, M.; Ichikawa, O.; Sasaki, K.; Miki, H.; Farahat, M.; Sawada, M. Selective Flotation of Chalcopyrite and Molybdenite with Plasma Pre-Treatment. *Miner. Eng.* **2014**, *66–68*, 102–111. [CrossRef]
23. King, R.P. Flotation. In *Modeling and Simulation of Mineral Processing Systems*; Butterworth Heinemann: Oxford, UK, 2001; pp. 289–350.
24. Xu, M. Modified Flotation Rate Constant and Selectivity Index. *Miner. Eng.* **1998**, *11*, 271–278. [CrossRef]
25. Bilal, M.; Ito, M.; Koike, K.; Hornn, V.; Ul Hassan, F.; Jeon, S.; Park, I.; Hiroyoshi, N. Effects of Coarse Chalcopyrite on Flotation Behavior of Fine Chalcopyrite. *Miner. Eng.* **2021**, *163*, 106776. [CrossRef]
26. Farahat, M.; Hirajima, T.; Sasaki, K.; Doi, K. Adhesion of Escherichia Coli onto Quartz, Hematite and Corundum: Extended DLVO Theory and Flotation Behavior. *Colloids Surf. B Biointerfaces* **2009**, *74*, 140–149. [CrossRef] [PubMed]
27. Aikawa, K.; Ito, M.; Kusano, A.; Park, I.; Oki, T.; Takahashi, T.; Furuya, H.; Hiroyoshi, N. Flotation of Seafloor Massive Sulfide Ores: Combination of Surface Cleaning and Deactivation of Lead-Activated Sphalerite to Improve the Separation Efficiency of Chalcopyrite and Sphalerite. *Metals* **2021**, *11*, 253. [CrossRef]

Article

Agglomeration–Flotation of Finely Ground Chalcopyrite Using Emulsified Oil Stabilized by Emulsifiers: Implications for Porphyry Copper Ore Flotation

Vothy Hornn [1,*], Mayumi Ito [2], Hiromasa Shimada [1], Carlito Baltazar Tabelin [3], Sanghee Jeon [2], Ilhwan Park [2] and Naoki Hiroyoshi [2]

1 Division of Sustainable Resources Engineering, Graduate School of Engineering, Hokkaido University, Sapporo 060-8628, Japan; simada.hiromasa.2525@gmail.com
2 Division of Sustainable Resources Engineering, Faculty of Engineering, Hokkaido University, Sapporo 060-8628, Japan; itomayu@eng.hokudai.ac.jp (M.I.); shjun1121@gmail.com (S.J.); i-park@eng.hokudai.ac.jp (I.P.); hiroyosi@eng.hokudai.ac.jp (N.H.)
3 School of Minerals and Energy Resources Engineering, The University of New South Wales, 2052 Sydney, NSW, Australia; c.tabelin@unsw.edu.au
* Correspondence: vothytalis1102@yahoo.com; Tel.: +81-011-706-6315

Received: 24 June 2020; Accepted: 7 July 2020; Published: 8 July 2020

Abstract: Flotation is the conventional method for processing porphyry copper deposits, one of the most economically important sources of copper (Cu) worldwide. The rapidly decreasing grade of this type of Cu ore in recent years, however, presents serious problems with fine particle recovery using conventional flotation circuits. This low recovery could be attributed to the low collision efficiency of fine particles and air bubbles during flotation. To improve collision efficiency and flotation recovery, agglomeration of finely ground chalcopyrite ($CuFeS_2$) (D_{50} = 3.5 μm) using emulsified oil stabilized by emulsifiers was elucidated in this study. Specifically, the effects of various types of anionic (sodium dodecyl sulfate (SDS), potassium amyl xanthate (KAX)), cationic (dodecyl amine acetate (DAA)), and non-ionic (polysorbate 20 (Tween 20)) emulsifiers on emulsified oil stability and agglomeration–flotation efficiency were investigated. When emulsifiers were added, the average size of agglomerates increased, resulting in higher Cu recovery during flotation. This dramatic improvement in flotation efficiency could be attributed to the smaller oil droplet size in emulsified oil and their higher stability in the presence of emulsifiers. The utilization of emulsifiers during agglomeration–flotation not only lowered the required agitation strength for agglomeration but also shortened the agglomeration time, both of which made the process easier to incorporate in existing flotation circuits.

Keywords: agglomeration; emulsified oil; emulsifiers; flotation; fine chalcopyrite

---

1. Introduction

Porphyry copper deposits are economically the most important source of copper (Cu) worldwide because of their wide distribution and enormous volume of minable ore materials. Some porphyry copper deposits have over 1 billion metric tons (t) of ore at > 0.5% of Cu [1], and many of these deposits typically contain other valuable metals aside from Cu, such as gold (Au) and molybdenum (Mo), making their exploitation economically attractive. For example, the largest open pit porphyry copper mine in the world—the Escondida mine in Chile—processes a porphyry copper deposit containing 0.25 g/t Au and 0.0062 wt% Mo [2]. The conventional technique used to recover valuable minerals from porphyry copper ores is flotation, a technique whereby fine particles of Cu-bearing minerals like chalcopyrite ($CuFeS_2$) are separated from gangue minerals and collected by bubbles after surface modifications using emulsifiers called collectors [3,4]. Flotation, including its concept of using bubbles to induce separation of materials with different surface wettability properties, is a popular and widely used technique not only in the mineral processing and coal cleaning/washing industries but also in electronic waste and plastic recycling industries [5–11].

Despite its popularity and widespread application, one major drawback of flotation is the dramatic drop in its efficiency when particle sizes become very small [12]. The low recovery of fine particles during flotation is attributed to the low collision probability between fine mineral particles and air bubbles [13–15]. Many studies have suggested ways to overcome the poor fine particle recovery during flotation using two approaches: (1) bubble size reduction, and (2) particle aggregation. Column flotation [16], microbubble flotation [17], electro-flotation [18], and dissolved air flotation [19] are some examples of bubble size reduction approaches, while particle aggregation strategies include shear flocculation [20], carrier flotation [21], polymer flocculation [22], and oil agglomeration [13].

All of these techniques have their own advantages and disadvantages, but from the perspective of economics, oil agglomeration—a method to increase the apparent particle size using oil as a "bridging" liquid under intense mixing—is considered as one of the most promising techniques for industrial-scale applications. Oil is relatively inexpensive, and the process could be easily integrated into existing flotation circuits. Although oil agglomeration has been extensively studied for coal cleaning [23,24], there is still one serious issue with this technique when applied to porphyry copper ores: the amount of oil required remains huge (up to 5–10% of total weight of coal for agglomeration-screening technique or around 0.1–1.0% for agglomeration–flotation technique) [25]. To address this problem, agglomeration–flotation using emulsified oil instead of simply adding oil during the agglomeration step has been proposed [26–28]. Emulsification of oil reduces the size of oil droplets, which lowers the amount of oil required for agglomeration dramatically.

With the decreasing grade and quality of porphyry copper deposits, flotation is increasingly becoming less efficient because of the smaller grain size of Cu-bearing minerals that require finer grinding for sufficient liberation [29]. As a result, fine particles cannot be recovered by conventional flotation, resulting in loss of valuable minerals. Trahar [15], for example, studied chalcopyrite–quartz flotation and reported that the recovery of chalcopyrite was lower when the median particle size decreased from 20 to 3 μm. In the previous work of the authors, agglomeration–flotation using emulsified oil instead of directly adding oil during the agglomeration step was conducted using finely ground chalcopyrite [4]. The results showed that the stability of emulsified oil was relatively short, so strong agitation strength is required to disperse oil droplets into water, maintain the small size of oil droplets, and facilitate particle agglomeration [4]. This strong agitation during agglomeration requires high energy, which make the process costly and problematic when integrated into actual flotation circuits.

One way to address this issue is to stabilize emulsified oil using emulsifiers during agglomeration–flotation. Emulsifiers are compounds that lower the surface tension between two immiscible liquids and have been reported to enhance not only the formation of small oil droplets but also stabilize them in aqueous medium [30]. To the best of our knowledge, agglomeration–flotation using emulsified oil stabilized by emulsifiers on finely ground chalcopyrite has never been studied before. In this study, the effects of emulsified oil stabilized by various types of emulsifiers on agglomeration and flotation of finely ground chalcopyrite were investigated.

2. Materials and Methods

2.1. Materials

The chalcopyrite sample used in this study was obtained from Copper Queen Mine, Arizona, USA. The sample was characterized by X-ray fluorescence (XRF) (EDXL300, Rigaku Corporation, Tokyo, Japan), X-ray powder diffraction (XRD, MultiFlex, Rigaku Corporation, Tokyo, Japan), and scanning electron microscopy (SEM, SM-IT200, JEOL Ltd., Tokyo, Japan).

The sample was crushed using a jaw crusher (1023-A, Yoshida Manufacturing Co., Ltd, Sapporo, Japan) and ground by a vibratory disc mill (RS100, Retsch Inc., Haan, Germany), then screened to obtain a size fraction of less than 75 µm. The ground sample (<75 µm) was ground again in the vibratory disc mill to obtain fine particles. The particle size distribution of the chalcopyrite sample used in this study was measured in water after ultrasonication using laser diffraction (Microtrac®MT3300SX, Nikkiso Co., Ltd., Tokyo, Japan).

Potassium amyl xanthate (KAX) (Tokyo Chemical Industry Co., Ltd., Tokyo, Japan), kerosene (Wako Pure Chemical Industries, Ltd., Osaka, Japan), methyl isobutyl carbinol (MIBC) (Tokyo Chemical Industry Co., Ltd., Tokyo, Japan), sodium dodecyl sulfate (SDS) (Tokyo Chemical Industry Co., Ltd., Japan), dodecyl amine acetate (DAA) (Tokyo Chemical Industry Co., Ltd., Tokyo, Japan), and polysorbate 20 (Tween20) (Tokyo Chemical Industry Co., Ltd., Tokyo Japan) were used in this study. The chemical structures of the four emulsifiers (SDS, KAX, DAA, and Tween 20) used in this study are shown in Figure 1.

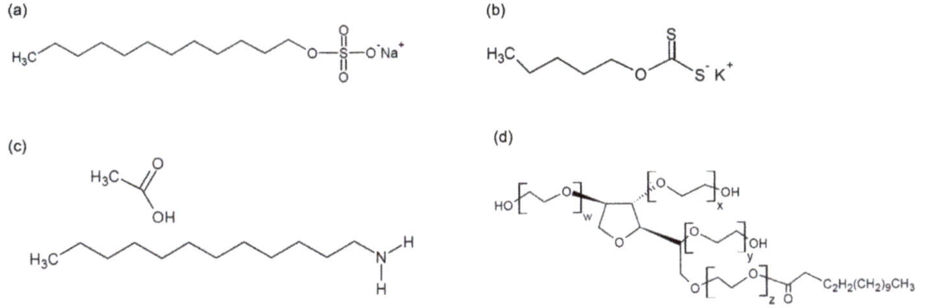

Figure 1. Chemical structures of emulsifiers: (**a**) SDS, (**b**) KAX, (**c**) DAA, and (**d**) Tween 20.

2.2. Methods

2.2.1. Preparation of Emulsified Oil with Emulsifiers

Anionic (SDS, KAX), cationic (DAA), and non-ionic (Tween 20) emulsifiers were used as emulsifying reagents. First, 10 mL of kerosene was mixed with 40 mL of deionized (DI) water, and then 0.1 g of the emulsifiers (SDS, KAX, DAA, or Tween 20) was added (i.e., concentration of emulsifiers, 2000 ppm). Emulsification of the mixture was then carried out before agglomeration using an ultrasonic homogenizer (ULTRA-TURRAX, IKA, Königswinter, Germany) for 60 s. After emulsification, oil droplet size was measured using laser diffraction.

2.2.2. Stability Tests of Emulsified Oil

Stability tests of emulsified oil containing different emulsifiers were carried out to evaluate the stability of the suspensions [31]. The emulsified oil after emulsification was equilibrated for 5 min until the bubble foams disappeared. Then, 50 mL of emulsified oil was centrifuged at 3500 rpm for 20 min. The volume ratios of oil and emulsion layer were calculated by measuring the thickness of oil layer.

2.2.3. KAX Conditioning and Agglomeration

Two conditions were evaluated during KAX conditioning and agglomeration as follows:

(a) Agglomeration at an agitation speed of 1000 rpm

i. KAX conditioning: Before agglomeration, 20 g of sample (D_{50} = 3.5 μm) was suspended in 400 mL of distilled water and then conditioned with the surface modifier, KAX (200 g/t), to improve hydrophobicity for 5 min at 1000 rpm in the flotation cell (FT-1000, Heiko, Tokyo, Japan).

ii. Agglomeration: After conditioning, emulsified oil (15 L kerosene/t sample, with and without surfactant) was added to the suspension, and agitation was carried out at 1000 rpm for 30 min in the flotation cell.

(b) Agglomeration at an agitation speed of 15,000 rpm

i. KAX conditioning: Before agglomeration, 20 g of sample (D_{50} = 3.5 μm) was suspended in 400 mL of distilled water and then conditioned with the surface modifier, KAX (200 g/t) for 5 min at 1000 rpm in the flotation cell.

ii. Agglomeration: After conditioning, the suspension was transferred to an agglomeration vessel (high speed mixer with a s-shaped impeller, SPB-600J, Cuisinart, Stamford, CT, USA; fixed rotation speed of 15,000 rpm), emulsified oil (15 L kerosene/t sample, with and without surfactant) was added, and agitation was carried out for 30 min. Suspension after agglomeration was then transferred to a 500 mL flotation cell.

The emulsified oil with emulsifiers was allowed to equilibrate for 5 min after emulsification prior to its use in the agglomeration step. In contrast, emulsified oil without any surfactant was used immediately because of the short stability of suspension. After agglomeration, the particle size distribution was measured by laser diffraction.

2.2.4. Flotation Tests

Flotation was carried out using a mechanical flotation machine (FT-1000, Heiko, Tokyo, Japan). MIBC (25 μL/L) was added as a frother, and the suspension was stirred for 3 min with an impeller speed of 1000 rpm. Air was then injected into the suspension at a flow rate of 1 L/min, and flotation was carried out (total flotation time of 10 min). Froths were collected at predetermined time intervals, and both froth products and tailings were weighed after drying at 105 °C for 24 h, and their chemical compositions were determined by XRF.

3. Results and Discussion

3.1. Characteristics of Chalcopyrite Samples

Chalcopyrite sample used in this study were characterized by XRF, XRD, SEM-EDX, and laser diffraction sizer. The chemical composition of the chalcopyrite sample (XRF) is shown in Table 1. Figure 2a,b shows the results of XRD and SEM-EDX analysis, respectively, of finely ground chalcopyrite. XRD analysis showed that the sample was mainly composed of chalcopyrite with minor amounts of actinolite ($Ca_2Mg_3Fe_2Si_8O_{22}(OH)_2$) and quartz. Figure 3 shows the particle size distributions of finely ground chalcopyrite (median particle diameter (D_{50}) of 3.5 μm), which was used in this study.

Table 1. Chemical composition of chalcopyrite sample.

Elements	Cu	Fe	S	Zn	Si	Ca
wt%	26	27	26	0.8	6	2

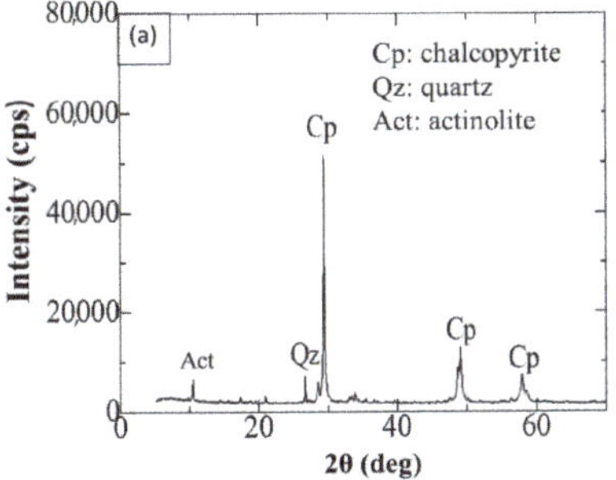

Figure 2. Chalcopyrite sample: (**a**) XRD patterns, and (**b**) SEM image.

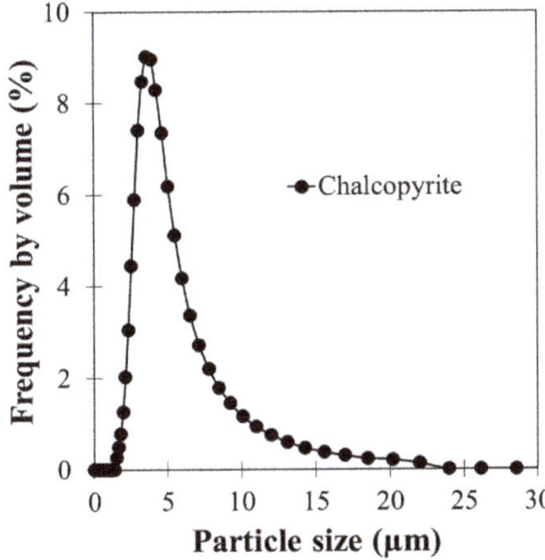

Figure 3. Particle size distribution of chalcopyrite sample.

3.2. Effects of Emulsifiers on Emulsified Oil Droplet Size and Stability

Anionic emulsifiers (SDS, KAX), cationic surfactant (DAA), and non-ionic surfactant (Tween 20) were used as emulsifying agents during the preparation of emulsified oil. These emulsifiers were chosen to represent three main types of flotation collector that have great potential to be used as emulsifying agents. Figure 4 shows the size distribution of oil droplets with and without addition of emulsifiers, and the results indicate that the mode size of oil droplets became smaller with emulsifiers. The smaller oil droplet generated in the presence of emulsifiers could be attributed to the reduction in interfacial tension of oil–water [32]. In the previous study of the authors [4], emulsified oil was added to the agglomeration vessel immediately because emulsified oil without surfactant was unstable. In a commercial plant, stability of emulsified oil is important to maintain the small size of oil droplets for oil agglomeration. To check the stability of the suspensions, stability tests were carried out. In this test, centrifugation was used to accelerate oil droplet coalescence, and because the thickness of oil layer relies on ease of coalescence and oil droplet size, this technique could be used to estimate the stability of emulsion. The volume ratios of oil and emulsion layers are shown in Figure 5. The volume of the oil layer after centrifugation was smaller when emulsifiers were present, indicating that these compounds were effective in stabilizing oil droplets and prolonging the usability of emulsified oil. Chen and Tao [32] defined the stability of an emulsion as the resistance by the dispersed oil droplets against coalescence and noted that emulsifying agents maintain the emulsion by forming a thin interfacial film between the two liquids that minimize contact, coalescence, and aggregation of the internal dispersed phase.

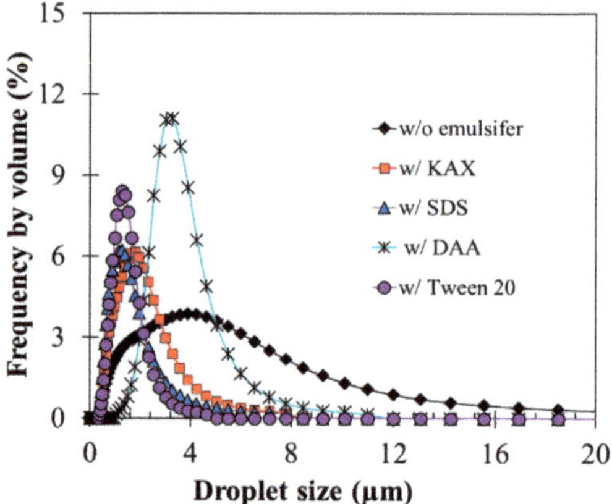

Figure 4. Size distribution of oil droplets after emulsification with and without addition of emulsifiers.

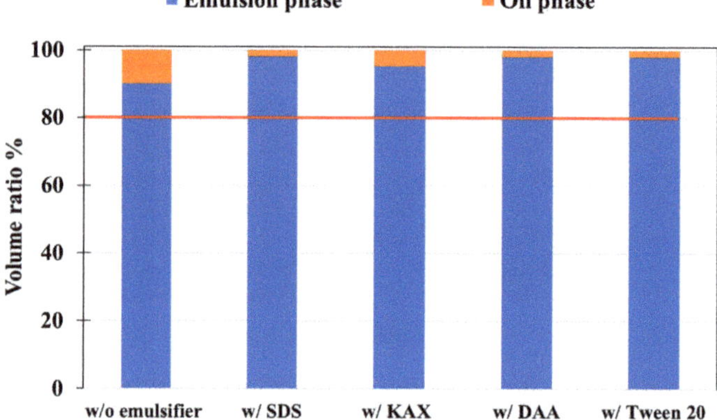

Figure 5. Volume ratio of oil and emulsion layers with and without emulsifiers after emulsification and centrifugation. Note that the red line indicates volume ratio without emulsification (80% water, 20% oil).

3.3. Effects of Emulsifiers on Agglomeration

In the previous section, SDS, KAX, DAA, and Tween 20 were shown to reduce the size of oil droplets and sustain the stability of emulsified oil. The effects of theses emulsifiers on agglomeration of chalcopyrite were investigated. Agglomeration in this study consisted of two stages: (1) conditioning in the flotation cell with 200 g/t of KAX, and (2) agglomeration using emulsified oil containing 2000 ppm of emulsifier (15 L kerosene/t). Figure 6 shows the apparent particle size distribution after agglomeration (1000 or 15,000 rpm for 30 min). The results showed that when SDS, KAX, and DAA were used, the mode size of agglomerate increased more than that without emulsifiers even at a high agitation speed (15,000 rpm). As shown in Figure 4, the droplet size of emulsified oil containing emulsifiers were relatively smaller than that without emulsifiers. When the size of oil droplets decreased, the number of droplets in the agitator increased. Because the frequency of collisions of the oil droplets and particles during agglomeration is proportional to the number of oil droplets, it is reasonable to deduce that the number of oil droplets attached to particle surfaces increased due to the higher collision probability between oil droplets and particles. Because of this, agglomerate size became bigger when emulsified oil containing emulsifiers were used.

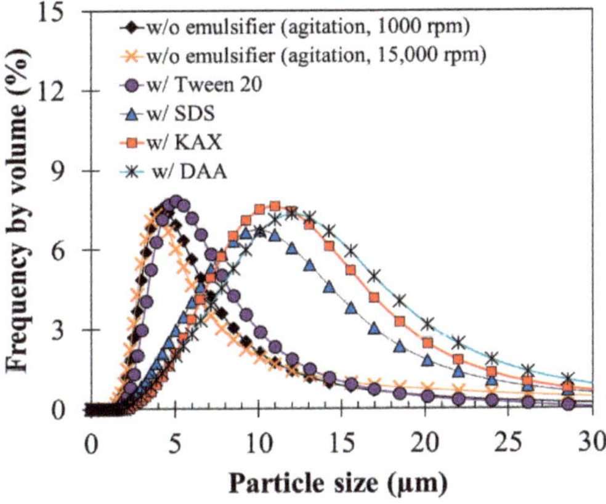

Figure 6. Size distribution after agglomeration (agglomeration conditions: 30 min; 1000 rpm with emulsifiers, 1000 and 15,000 rpm without emulsifiers).

When Tween 20 was used, the mode size of agglomerates was similar to that without surfactant. Agglomeration kinetics and agglomerate size are mainly determined by three processes: (1) particle–particle collision, (2) particle–particle attraction, and (3) decomposition of agglomerate [33]. As shown in Figure 4, the size of droplets with Tween 20 were similar to those with other types of emulsifiers, indicating that the probability of collision between particles and oil droplets may be similar. However, the probability of attraction of particles and oil droplets stabilized by Tween 20 may be lower than those with the other emulsifiers.

3.4. Effects of Emulsifiers on Flotation

In this flotation test, agglomeration using emulsified oil with and without emulsifiers was carried out, and flotation of the agglomerates was conducted. Figure 7 shows the Cu recovery with time (agglomeration conditions: 30 min; 1000 rpm with emulsifiers, 1000 and 15,000 rpm without emulsifiers). Without emulsifiers, Cu recovery with agglomeration using high agitation strength (15,000 rpm) was higher than that with low agitation strength (1000 rpm), indicating that intense agitation is required for agglomeration–flotation without emulsifiers due to the low stability of emulsified oil.

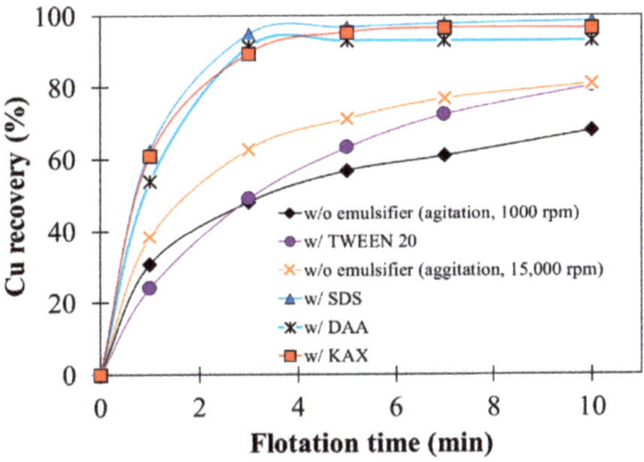

Figure 7. Cu recovery with time (agglomeration conditions: 30 min; 1000 rpm with emulsifiers, 1000 and 15,000 rpm without emulsifiers).

In the case of agglomeration–flotation with Tween 20, Cu recovery showed a similar trend with that without emulsifiers. This result is in-line with the result of size distribution of agglomerate; that is, the size of agglomerate using Tween 20 was almost the same as that without emulsifier. The recovery rate of chalcopyrite treated with Tween 20 was lower than those treated with other emulsifiers. Cu recovery rate was affected by the size of agglomerate, as described before and agglomerate size may have been influenced by the affinity between the oil-emulsion and the KAX-adsorbed chalcopyrite surface (KAX was used as the surface modifier to increase the hydrophobicity). Interaction forces between hydrophobic minerals surface and oil emulsion may be determined by the balance of hydrophobic interaction, van der Waals attraction, and steric repulsion of the structure of the hydrophilic head of surfactants. Hydrophilic lipophilic balance (HLB), an indicator of the balance of hydrophilic and hydrophobic groups of surfactants, strongly affects the properties of emulsions, including oil droplet size. HLB values of Tween 20 and SDS are similar, resulting in similar oil droplets sizes [34,35]. However, the size of agglomerate in Tween 20-stabilized emulsion was smaller than that with SDS, indicating that interactions between chalcopyrite and oil droplets with Tween 20 or SDS were different. One possible explanation for the low efficiency of Tween 20 in agglomerating finely ground chalcopyrite is its complex structure consisting of numerous types of hydrophilic heads compared to SDS, as shown in Figure 1. In other words, Tween 20, having hydrophilic heads, may have lowered the hydrophobicity of the oil-emulsion, thereby decreasing the affinity between oil droplets and chalcopyrite.

In the case of agglomeration–flotation with SDS, DAA, and KAX, Cu recovery with agglomeration using these emulsifiers was higher than that in agglomeration with intense agitation strength (15,000 rpm) without emulsifiers. These results indicate that less intense agitation strength (1000 rpm) was sufficient for agglomeration–flotation when emulsifiers were present because of the smaller droplet size and better stability of emulsified oil. This means that when emulsifiers are used, specialized equipment with high agitation strengths are not required, and the process could be easily integrated into existing flotation circuits.

Figure 8 shows the Cu recovery of agglomerate with SDS at agglomeration times of 15 and 30 min, and the results showed that Cu recovery was very similar, indicating that agglomeration for 15 min was enough to achieve high Cu recovery. In other words, utilization of emulsifiers–stabilized oil emulsion could shorten the agglomeration time.

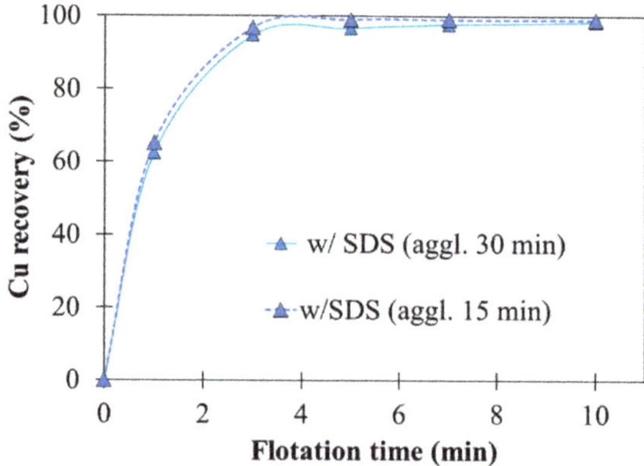

Figure 8. Cu recovery of agglomerate with SDS at agglomeration time of 15 and 30 min.

4. Conclusions

The following conclusions can be drawn from the agglomeration–flotation study using emulsified oil with emulsifiers:

- Addition of emulsifiers (e.g., SDS, KAX, DAA, and Tween 20) reduced the size of oil droplets and maintained the stability of emulsified oil longer.
- Addition of emulsifiers (SDS, KAX, and DAA) to emulsified oil contributed to the formation of bigger agglomerates, thus improving Cu recovery by agglomeration–flotation even at low agitation strength and shorter agglomeration time.
- When emulsified oil with emulsifiers is used, specialized equipment with higher agitation strength during agglomeration is not required, and thus the process could be easily integrated into existing flotation circuits.

Author Contributions: V.H., M.I., S.J., I.P., and N.H. conceived and designed the experiments; V.H. and H.S. performed the experiments and analyzed the data; V.H. wrote the paper. M.I., C.B.T., S.J., and I.P. reviewed and edited the paper. All authors have read and agreed to the published version of the manuscript.

Funding: This research received no external funding.

Conflicts of Interest: The authors declare no conflict of interest.

References

1. Brown, T.J.; Bide, T.; Walters, A.S.; Idoine, N.E.; Shaw, R.A.; Hannis, S.D.; Lusty, P.A.J.; Kendall, R.; MacKenzie, A.C. *World Mineral Production 2005-09*; British Geological Survey: Nottingham, UK, 2011; pp. 29–31.
2. Richards, J.P.; Boyce, A.J.; Pringle, M.S. Geologic Evolution of the Escondida Area, Northern Chile: A Model for Spatial and Temporal Localization of Porphyry Cu Mineralization. *Econ. Geol. Bull. Soc. Econ. Geol.* **2001**, *96*, 271–305. [CrossRef]
3. Aikawa, K.; Ito, M.; Segawa, T.; Jeon, S.; Park, I.; Tabelin, C.B.; Hiroyoshi, N. Depression of Lead-Activated Sphalerite by Pyrite via Galvanic Interactions: Implications to the Selective Flotation of Complex Sulfide Ores. *Miner. Eng.* **2020**, *152*, 106367. [CrossRef]
4. Hornn, V.; Ito, M.; Shimada, H.; Tabelin, C.B.; Jeon, S.; Park, I.; Hiroyoshi, N. Agglomeration-Flotation of Finely Ground Chalcopyrite and Quartz: Effects of Agitation Strength during Agglomeration Using Emulsified Oil on Chalcopyrite. *Minerals* **2020**, *10*, 380. [CrossRef]
5. Ito, M.; Saito, A.; Murase, N.; Phengsaart, T.; Kimura, S.; Tabelin, C.B.; Hiroyoshi, N. Development of Suitable Product Recovery Systems of Continuous Hybrid Jig for Plastic-Plastic Separation. *Miner. Eng.* **2019**, *141*, 105839. [CrossRef]
6. Ito, M.; Takeuchi, M.; Saito, A.; Murase, N.; Phengsaart, T.; Tabelin, C.B.; Hiroyoshi, N.; Tsunekawa, M. Improvement of Hybrid Jig Separation Efficiency Using Wetting Agents for the Recycling of Mixed-Plastic Wastes. *J. Mater. Cycles Waste Manag.* **2019**, *21*, 1376–1383. [CrossRef]
7. Jeon, S.; Ito, M.; Tabelin, C.B.; Pongsumrankul, R.; Kitajima, N.; Park, I.; Hiroyoshi, N. Gold Recovery from Shredder Light Fraction of E-Waste Recycling Plant by Flotation-Ammonium Thiosulfate Leaching. *Waste Manag.* **2018**, *77*, 195–202. [CrossRef]
8. Jeon, S.; Ito, M.; Tabelin, C.B.; Pongsumrankul, R.; Tanaka, S.; Kitajima, N.; Saito, A.; Park, I.; Hiroyoshi, N. A Physical Separation Scheme to Improve Ammonium Thiosulfate Leaching of Gold by Separation of Base Metals in Crushed Mobile Phones. *Miner. Eng.* **2019**, *138*, 168–177. [CrossRef]
9. Phengsaart, T.; Ito, M.; Azuma, A.; Tabelin, C.B.; Hiroyoshi, N. Jig Separation of Crushed Plastics: The Effects of Particle Geometry on Separation Efficiency. *J. Mater. Cycles Waste Manag.* **2020**, *22*, 787–800. [CrossRef]
10. Phengsaart, T.; Ito, M.; Hamaya, N.; Tabelin, C.B.; Hiroyoshi, N. Improvement of Jig Efficiency by Shape Separation, and a Novel Method to Estimate the Separation Efficiency of Metal Wires in Crushed Electronic Wastes Using Bending Behavior and "Entanglement Factor". *Miner. Eng.* **2018**, *129*, 54–62. [CrossRef]
11. Seng, S.; Tabelin, C.B.; Makino, Y.; Chea, M.; Phengsaart, T.; Park, I.; Hiroyoshi, N.; Ito, M. Improvement of Flotation and Suppression of Pyrite Oxidation Using Phosphate-Enhanced Galvanic Microencapsulation (GME) in a Ball Mill with Steel Ball Media. *Miner. Eng.* **2019**, *143*, 105931. [CrossRef]

12. Jiangang, F.; Kaida, C.; Hui, W.; Chao, G.; Wei, L. Recovering Molybdenite from Ultrafine Waste Tailings by Oil Agglomerate Flotation. *Miner. Eng.* **2012**, *39*, 133–139. [CrossRef]

13. Dai, Z.; Fornasiero, D.; Ralston, J. Particle–Bubble Collision Models—a Review. *Adv. Colloid Interface Sci.* **2000**, *85*, 231–256. [CrossRef]

14. Miettinen, T.; Ralston, J.; Fornasiero, D. The Limits of Fine Particle Flotation. *Miner. Eng.* **2010**, *23*, 420–437. [CrossRef]

15. Trahar, W.J.; Warren, L.J. The Flotability of Very Fine Particles—A Review. *Int. J. Miner. Process.* **1976**, *3*, 103–131. [CrossRef]

16. Finch, J.A. Column Flotation: A Selected Review— Part IV: Novel Flotation Devices. *Miner. Eng.* **1995**, *8*, 587–602. [CrossRef]

17. Yoon, R.H. Microbubble Flotation. *Miner. Eng.* **1993**, *6*, 619–630. [CrossRef]

18. Bhaskar Raju, G.; Khangaonkar, P.R. Electro-Flotation of Chalcopyrite Fines. *Int. J. Miner. Process.* **1982**, *9*, 133–143. [CrossRef]

19. Rodrigues, R.T.; Rubio, J. DAF–Dissolved Air Flotation: Potential Applications in the Mining and Mineral Processing Industry. *Int. J. Miner. Process.* **2007**, *82*, 1–13. [CrossRef]

20. Warren, L.J. Shear-Flocculation of Ultrafine Scheelite in Sodium Oleate Solutions. *J. Colloid Interface Sci.* **1975**, *50*, 307–318. [CrossRef]

21. Rubio, J.; Hoberg, H. The Process of Separation of Fine Mineral Particles by Flotation with Hydrophobic Polymeric Carrier. *Int. J. Miner. Process.* **1993**, *37*, 109–122. [CrossRef]

22. Sresty, G.C.; Somasundaran, P. Selective Flocculation of Synthetic Mineral Mixtures Using Modified Polymers. *Int. J. Miner. Process.* **1980**, *6*, 303–320. [CrossRef]

23. Alonso, M.I.; Valdés, A.F.; Martínez-Tarazona, R.M.; Garcia, A.B. Coal Recovery from Coal Fines Cleaning Wastes by Agglomeration with Vegetable Oils: Effects of Oil Type and Concentration. *Fuel* **1999**, *78*, 753–759. [CrossRef]

24. Slaghuis, J.H.; Ferreira, L.C. Selective Spherical Agglomeration of Coal. *Fuel* **1987**, *66*, 1427–1430. [CrossRef]

25. Laskowski, J.S.; Liu, Q.; Bolin, N.J. Flotation of Sulphide Minerals 1990 Polysaccharides in Flotation of Sulphides. Part I. Adsorption of Polysaccharides onto Mineral Surfaces. *Int. J. Miner. Process.* **1991**, *33*, 223–234. [CrossRef]

26. Bensley, C.N.; Swanson, A.R.; Nicol, S.K. The Effect of Emulsification on the Selective Agglomeration of Fine Coal. *Int. J. Miner. Process.* **1977**, *4*, 173–184. [CrossRef]

27. Van Netten, K.; Moreno-Atanasio, R.; Galvin, K.P. Fine Particle Beneficiation through Selective Agglomeration with an Emulsion Binder. *Ind. Eng. Chem. Res.* **2014**, *53*, 15747–15754. [CrossRef]

28. Sahinoglu, E.; Uslu, T. Use of Ultrasonic Emulsification in Oil Agglomeration for Coal Cleaning. *Fuel* **2013**, *113*, 719–725. [CrossRef]

29. Johnson, N.W. Liberated 0–10μm Particles from Sulphide Ores, Their Production and Separation—Recent Developments and Future Needs. *Miner. Eng.* **2006**, *19*, 666–674. [CrossRef]

30. Stang, M.; Karbstein, H.; Schubert, H. Adsorption Kinetics of Emulsifiers at Oil—Water Interfaces and Their Effect on Mechanical Emulsification. *Chem. Eng. Process. Process Intensif.* **1994**, *33*, 307–311. [CrossRef]

31. Chen, G.; Tao, D. An Experimental Study of Stability of Oil–Water Emulsion. *Fuel Process. Technol.* **2005**, *86*, 499–508. [CrossRef]

32. Oh, S.G.; Shah, D.O. Effect of counterions on the interfacial tension and emulsion droplet size in the oil/water/dodecyl sulfate system. *J. Phys. Chem.* **1993**, *97*, 284–286. [CrossRef]

33. Darelius, A.; Rasmuson, A.; Björn, I.N.; Folestad, S. High shear wet granulation modelling—a mechanistic approach using population balances. *Powder Technol.* **2005**, *160*, 209–218. [CrossRef]

34. Kunieda, H.; Hanno, K.; Yamaguchi, S.; Shinoda, K. The Three-Phase Behavior of a Brine/Ionic Surfactant/Nonionic Surfactant/Oil System: Evaluation of the Hydrophile-Lipophile Balance (HLB) of Ionic Surfactant. *J. Colloid Interface Sci.* **1985**, *107*, 129–137. [CrossRef]

35. Kunieda, H.; Ishikawa, N. Evaluation of the Hydrophile-Lipophile Balance (HLB) of Nonionic Surfactants. II. Commercial-Surfactant Systems. *J. Colloid Interface Sci.* **1985**, *107*, 122–128. [CrossRef]

Article

Flotation of Seafloor Massive Sulfide Ores: Combination of Surface Cleaning and Deactivation of Lead-Activated Sphalerite to Improve the Separation Efficiency of Chalcopyrite and Sphalerite

Kosei Aikawa [1,*], Mayumi Ito [2], Atsuhiro Kusano [1], Ilhwan Park [2], Tatsuya Oki [3], Tatsuru Takahashi [4], Hisatoshi Furuya [5] and Naoki Hiroyoshi [2]

[1] Division of Sustainable Resources Engineering, Graduate School of Engineering, Hokkaido University, Sapporo 060-8628, Japan; kusano.atsuhiro.z2@elms.hokudai.ac.jp

[2] Division of Sustainable Resources Engineering, Faculty of Engineering, Hokkaido University, Sapporo 060-8628, Japan; itomayu@eng.hokudai.ac.jp (M.I.); i-park@eng.hokudai.ac.jp (I.P.); hiroyosi@eng.hokudai.ac.jp (N.H.)

[3] Environmental Management Research Institute, National Institute of Advanced Industrial Science and Technology (AIST), Tsukuba 305-8569, Japan; t-oki@aist.go.jp

[4] Seafloor Mineral Resources Research & Development Division, Seafloor Mineral Resources Department, Japan Oil, Gas and Metals National Corporation (JOGMEC), Tokyo 105-0001, Japan; takahashi-tatsuru@jogmec.go.jp

[5] Metals Technology Division, Metals Development Department, Japan Oil, Gas and Metals National Corporation (JOGMEC), Tokyo 105-0001, Japan; furuya-hisatoshi@jogmec.go.jp

* Correspondence: k88a28_tennis@eis.hokudai.ac.jp; Tel.: +81-11-706-6315

Citation: Aikawa, K.; Ito, M.; Kusano, A.; Park, I.; Oki, T.; Takahashi, T.; Furuya, H.; Hiroyoshi, N. Flotation of Seafloor Massive Sulfide Ores: Combination of Surface Cleaning and Deactivation of Lead-Activated Sphalerite to Improve the Separation Efficiency of Chalcopyrite and Sphalerite. *Metals* **2021**, *11*, 253. https://doi.org/10.3390/met11020253

Academic Editor: Fernando Castro
Received: 28 December 2020
Accepted: 29 January 2021
Published: 2 February 2021

Publisher's Note: MDPI stays neutral with regard to jurisdictional claims in published maps and institutional affiliations.

Abstract: The purpose of this study is to propose the flotation procedure of seafloor massive sulfide (SMS) ores to separate chalcopyrite and galena as froth and sphalerite, pyrite, and other gangue minerals as tailings, which is currently facing difficulties due to the presence of water-soluble compounds. The obtained SMS ore sample contains $CuFeS_2$, ZnS, FeS_2, SiO_2, and $BaSO_4$ in addition to PbS and $PbSO_4$ as Pb minerals. Soluble compounds releasing Pb, Zn^{2+}, Pb^{2+}, and $Fe^{2+/3+}$ are also contained. When anglesite co-exists, lead activation of sphalerite occurred, and thus sphalerite was recovered together with chalcopyrite as froth. To remove soluble compounds (e.g., anglesite) that have detrimental effects on the separation efficiency of chalcopyrite and sphalerite, surface cleaning pretreatment using ethylene diamine tetra acetic acid (EDTA) was applied before flotation. Although most of anglesite were removed and the recovery of chalcopyrite was improved from 19% to 81% at 20 g/t potassium amyl xanthate (KAX) after EDTA washing, the floatability of sphalerite was not suppressed. When zinc sulfate was used as a depressant for sphalerite after EDTA washing, the separation efficiency of chalcopyrite and sphalerite was improved due to deactivation of lead-activated sphalerite by zinc sulfate. The proposed flotation procedure of SMS ores—a combination of surface cleaning with EDTA to remove anglesite and the depression of lead-activated sphalerite by using zinc sulfate—could achieve the highest separation efficiency of chalcopyrite and sphalerite; that is, at 200 g/t KAX, the recoveries of chalcopyrite and sphalerite were 86% and 17%, respectively.

Keywords: seafloor massive sulfide; flotation; lead-activated sphalerite; anglesite; EDTA

1. Introduction

Seafloor massive sulfide (SMS) deposits, also referred to as submarine hydrothermal polymetallic sulfide deposits, have gained increasing attention as new metal resources because they consist of various forms of polymetallic sulfide including Cu, Pb, Zn, Au, Ag, etc. SMS deposits were found in a variety of volcanic and tectonic settings on the modern ocean floor in worldwide [1,2]. A number of scholars have studied SMS deposits, examples of which include formation mechanisms [3–5], exploration [6–8], mining methods [9–12],

65

and environmental impacts [13–15]; however, only a few studies on mineral processing of SMS ores have been carried out [16,17].

In Japan, the program for the development of SMS deposits around Japan has been conducted by the Ministry of Economic, Trade and Industry (METI) and Japan Oil, Gas and Metals National Corporation (JOGMEC). Not only seafloor mining, but also mineral processing and extractive metallurgy of SMS ores have been studied. Masuda (2011) [18] reported that SMS resembles the Kuroko on land in terms of deposit origin, which implies that similar processing and refining technologies applied to Kuroko can also be utilized for SMS ores. However, the previous studies on mineral processing of SMS ores conducted by JOGMEC indicated that mineral processing behaviors of the SMS ores are quite different compared to those of Kuroko [17]. METI and JOGMEC (2018) [16] reported distinct features of SMS ores obtained from around Japan: (1) lead minerals in the SMS ores are mostly present as anglesite ($PbSO_4$) with a minor amount of galena (PbS), and (2) metal ions are highly dissolved from the SMS ores (e.g., [Zn^{2+}], 1700–3000 ppm). Moreover, some researchers reported various metal ions like Cu^{2+}, Pb^{2+}, and Zn^{2+} are released from the SMS ores obtained from the Trans-Atlantic Geotraverse active mound on the Mid-Atlantic Ridge [10] and the Izena Hole in the middle Okinawa Trough, Japan [11]. These distinct features of SMS ores around Japan—the presence of anglesite and the release of metal ions—would make mineral processing of SMS ores composed of Cu-Pb-Zn sulfide minerals complicated.

Cu-Pb-Zn sulfide minerals are generally processed via two flotation stages whereby Cu- and Pb-sulfide minerals are first recovered, followed by floating Zn-sulfide minerals [19]. When anglesite is contained in Cu-Pb-Zn sulfide minerals, it is readily dissolved and releases Pb^{2+} that activates sphalerite via the formation of PbS-like compounds on sphalerite surface, as explained in Equation (1). Activation of sphalerite by Pb^{2+} is known to increase the sphalerite floatability because the PbS-like compound has a higher affinity with xanthate than sphalerite [20–23].

$$ZnS(s) + Pb^{2+} = PbS(s) + Zn^{2+} \tag{1}$$

Thus, activation of sphalerite by Pb^{2+} is an unwelcomed side reaction because it dramatically limits the separation of Cu-Pb-Zn sulfide minerals. To improve their separation efficiency in the presence of anglesite, depression of lead-activated sphalerite is necessary. In practice, zinc sulfate and sulfoxy reagents (e.g., sulfite, metabisulfite, and bisulfate) are used as depressants for sphalerite [24] and, moreover, the former can also be used for the depression of lead-activated sphalerite [25,26]. The aim of this study is to propose a flotation procedure of SMS ores that is facing difficulty of separating chalcopyrite and sphalerite due to the presence of water-soluble compounds like anglesite. To improve the separation efficiency of chalcopyrite and sphalerite, depression effects of sodium sulfite and zinc sulfate on sphalerite floatability in the flotation of SMS ores were examined. In addition, a chemical pretreatment using ethylene diamine tetra acetic acid (EDTA) to remove anglesite and other soluble species (e.g., oxidation products) having potentials to affect flotation process was investigated. Finally, a flotation procedure of SMS ores to separate chalcopyrite and sphalerite in the presence of soluble compounds like anglesite using a combination of surface cleaning with EDTA and depression of lead-activated sphalerite by zinc sulfate was proposed.

2. Materials and Methods

2.1. Samples

A submarine hydrothermal polymetallic sulfide ore sample (named as sample A) obtained from around Japan and provided by JOGMEC, and seven types of minerals were used in this study: chalcopyrite ($CuFeS_2$, Copper Queen Mine, Cochise County, AZ, USA), sphalerite (ZnS, Kamioka Mine, Hida, Japan), galena (PbS, Beni Tadjit, Figuig, Morocco), anglesite ($PbSO_4$, Puit 9 Touissit, Oujda,, Morocco), pyrite (FeS_2, Huanzala Mine, Huanuco, Peru), quartz (SiO_2, 99% purity, Wako Pure Chemical Industries Co., Ltd., Tokyo, Japan), and barite ($BaSO_4$, Jungcheon Changdo Mine, Kimhwa County, South Korea). The mixtures of chalcopyrite, sphalerite, pyrite, quartz, and barite with galena or anglesite were used as model samples for flotation experiments. Sample A and the above-mentioned mineral samples were characterized using X-ray fluorescence spectroscopy (XRF, EDXL300, Rigaku Corporation, Tokyo, Japan) and X-ray powder diffraction (XRD, MultiFlex, Rigaku Corporation, Tokyo, Japan), and the chemical and mineralogical compositions of these samples are summarized in Table 1 and shown in Figure 1, respectively. The XRD pattern of the submarine hydrothermal polymetallic sulfide ore sample (sample A) shows that it contains chalcopyrite, sphalerite, galena, anglesite, barite, pyrite, and quartz (Figure 1a).

Table 1. Chemical composition of sample A based on X-ray fluorescence spectroscopy (XRF).

Sample	Mass Fraction (%)						
	Cu	**Zn**	**Pb**	**Fe**	**S**	**Si**	**Ba**
Sample A	7.4	13.6	7.1	24.5	35.7	3.5	1.5
Chalcopyrite	24.5	0.7	-	34.1	26.0	8.8	-
Sphalerite	-	66.5	0.1	3.5	24.8	2.7	-
Galena	-	-	84.8	-	8.3	1.5	-
Anglesite	1.3	0.7	88.0	0.2	7.6	0.7	-
Pyrite	-	-	-	42.3	52.5	1.0	-
Barite	-	-	-	-	17.9	0.3	67.8

The samples were ground by using a vibratory disc mill (RS 100, Retsch Inc., Haan, Germany) and were screened to obtain a size fraction of -75 μm. For the flotation experiments, potassium amyl xanthate (KAX, Tokyo Chemical Industry Co., Ltd., Tokyo, Japan) as a collector, Methyl Isobutyl Carbinol (MIBC, Tokyo Chemical Industry Co., Ltd., Tokyo, Japan) as a frother, and sodium sulfite (Na_2SO_3, Wako Pure Chemical Industries, Ltd., Osaka, Japan) and zinc sulfate ($ZnSO_4$, Wako Pure Chemical Industries Ltd., Osaka, Japan) as depressants were used. Sodium hydroxide (NaOH, Wako Pure Chemical Industries Ltd., Osaka, Japan) and sulfuric acid (H_2SO_4, Wako Pure Chemical Industries, Ltd., Osaka, Japan) were used as pH adjusters. For surface cleaning to remove the oxidation products present on mineral surface and/or contained in sample A, ethylene diamine tetra acetic acid (EDTA, Wako Pure Chemical Industries, Ltd., Osaka, Japan) was used [27,28].

2.2. Experimental Methods

2.2.1. Flotation

Prior to flotation tests, samples were deslimed by the following procedure: (1) a 20 g sample was added into 300 mL distilled water and then ultrasonication using high-speed switching oscillation between 24 kHz and 31 kHz (W-113 MK-II, Honda Electronics Co., Ltd., Toyohashi, Japan) was carried out for 1 min, (2) the suspension was allowed to be settled down for 5 min, and then the supernatant was removed. These treatments were done in triplicate, and the settled portion was used for flotation experiments.

An agitator-type flotation machine (FT-1000, Heiko-Seisakusyo, Tokyo, Japan) equipped with a 400-mL flotation cell was used and flotation experiments were conducted under the following conditions: pH, 6.5; temperature, 25 °C; pulp density, 5%; impeller speed, 1000 rpm; air flow rate, 1 L/min. After flotation, froth and tailing products were dried in an oven at 105 °C for 24 h and analyzed by XRF to determine the recovery of Cu, Zn, Pb, Fe, Si, and Ba. Flotation experiments were carried out based on the flowchart, as illustrated in Figure 2.

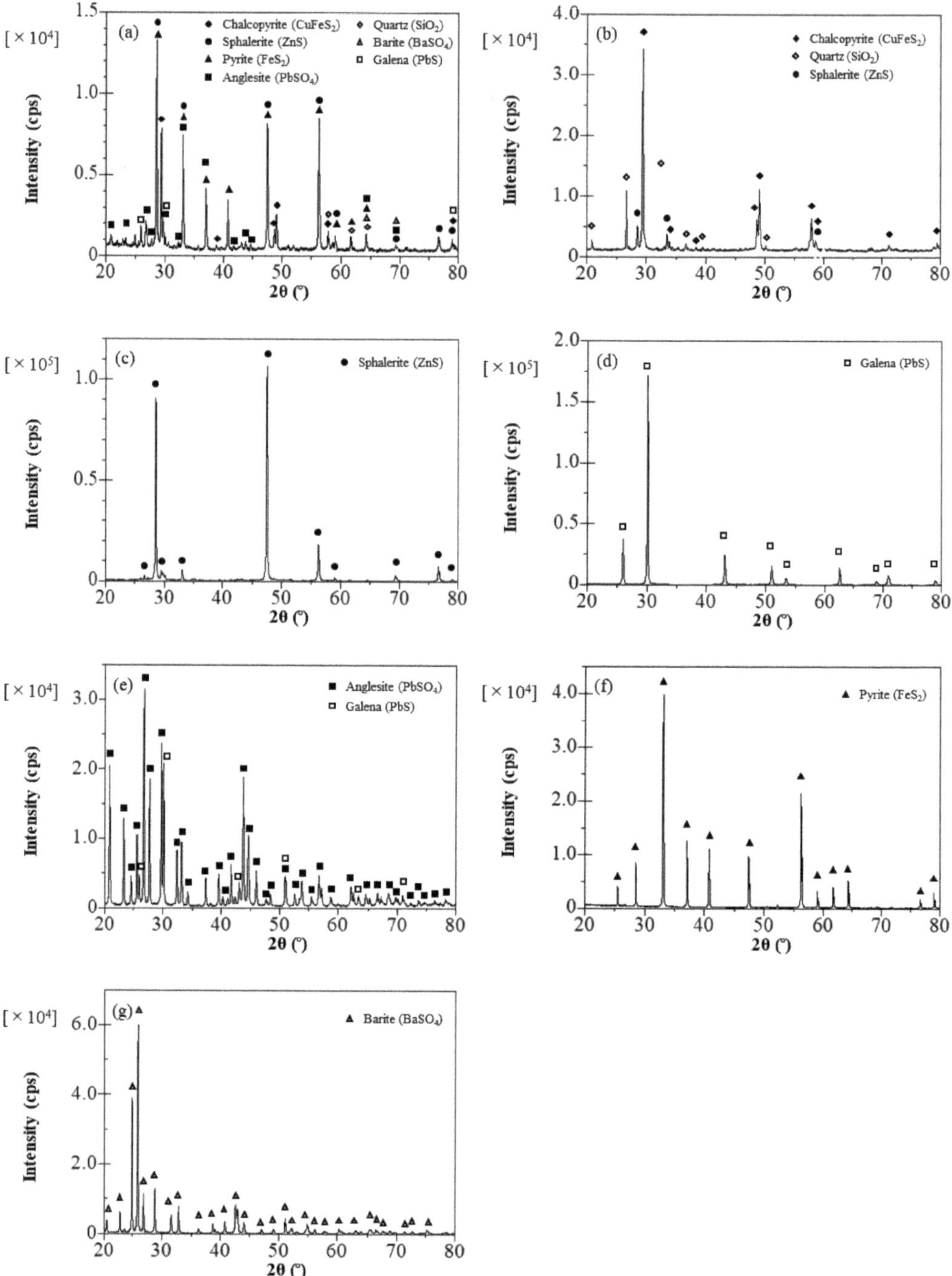

Figure 1. The X-ray powder diffraction (XRD) patterns of (**a**) sample A, (**b**) chalcopyrite, (**c**) sphalerite, (**d**) galena, (**e**) anglesite, (**f**) pyrite, and (**g**) barite. Note differences in the scale of the y-axes.

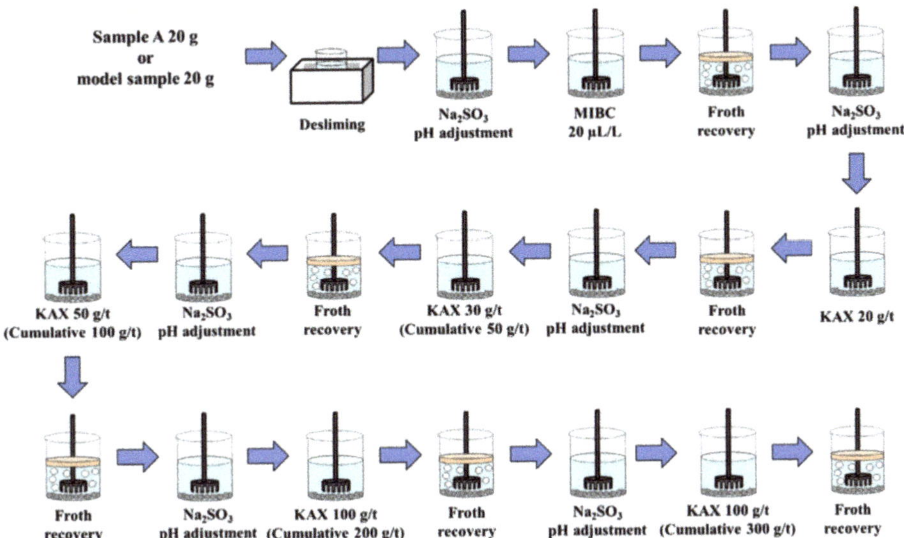

Figure 2. Flotation procedure with sequential addition of collector (KAX).

2.2.2. Surface Cleaning of Sample A

Surface cleaning with EDTA: (1) a 20 g sample was added into 300 mL solution containing EDTA (25 g/L) and then ultrasonicated for 1 min, (2) the supernatant of the suspension was removed after settling for 5 min. These treatments were done in triplicate and flotation experiments were conducted immediately after this surface cleaning.

2.2.3. Leachability Test of Sample A with DI Water

To check the amounts of soluble species in sample A, a leachability test was carried out. To achieve this, 0.4 g of sample A and 40 mL distilled water (i.e., pulp density: 1%) were added into a 50-mL centrifuge tube and shaken by a roller shaker (MIX ROTOR VMR-5R, AS ONE Co., Ltd., Osaka, Japan) at 100 rpm for 10 min. Afterward, the leachate was collected by filtration using 0.2 μm syringe-driven membrane filters and immediately analyzed by an inductively coupled plasma atomic emission spectrometer (ICP-AES, ICPE 9820, Shimadzu Corporation, Kyoto, Japan) (margin of error = $\pm 2\%$) to measure the concentration of Cu^{2+}, Pb^{2+}, Zn^{2+}, and $Fe^{2+/3+}$ released from sample A.

3. Results and Discussion

3.1. Effect of Sodium Sulfite on the Separation of Chalcopyrite and Sphalerite in the Flotation of Sample A

Figure 3 shows the flotation results of sample A with various dosages of sodium sulfite (0, 5, or 20 kg/t) to evaluate its suppressive effect on the floatability of pyrite as well as sphalerite. The floatability of pyrite was apparently suppressed as Na_2SO_3 dosage increased; that is, about 45% of pyrite was recovered at 100 g/t KAX in the absence of Na_2SO_3 (Figure 3a), but it decreased to ~30% with 5 kg/t Na_2SO_3 and ~22% with 20 kg/t Na_2SO_3 (Figure 3b,c). However, sphalerite was recovered as froth together with chalcopyrite, irrespective of the amount of depressant added. In the case of Pb minerals (i.e., anglesite and galena), their recovery was low at around 30–40%, suggesting that anglesite is most likely the main Pb mineral in sample A. As mentioned earlier, the presence of anglesite can dramatically change the flotation behavior of sphalerite because of Pb^{2+} released from anglesite that activates sphalerite.

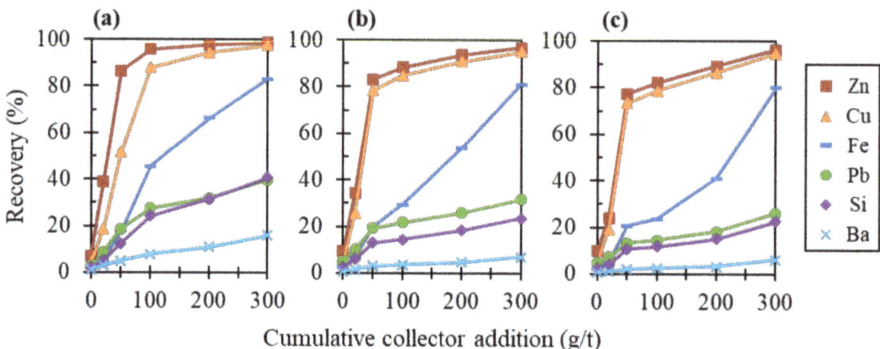

Figure 3. Flotation results of sample A with sodium sulfite (**a**) 0 kg/t, (**b**) 5 kg/t, and (**c**) 20 kg/t.

To clarify the effects of Pb minerals on the floatability of sphalerite, flotation tests were conducted using model samples, prepared based on the actual mineralogical composition of sample A obtained by norm calculation (Table 2).

Table 2. Norm calculation results of sample A.

		Mass Fraction (%)			
$CuFeS_2$	ZnS	PbS	FeS_2	SiO_2	$BaSO_4$
21.7	20.6	8.4	39.2	7.5	2.6

Specifically, two types of model samples (PbS-type and $PbSO_4$-type) were prepared considering PbS or $PbSO_4$ as the only Pb mineral. Flotation experiments using the PbS-type or the $PbSO_4$-type model sample were carried out with 5 kg/t Na_2SO_3. As shown in Figure 4, chalcopyrite and galena were floated first followed by sphalerite in the flotation of the PbS-type model sample, whereas sphalerite was floated together with chalcopyrite in the flotation of the $PbSO_4$-type model sample. In general, Cu-Pb-Zn sulfide ores are processed via two flotation stages whereby Cu- and Pb-sulfide minerals are first recovered, followed by Zn-sulfide minerals. The flotation result of the PbS-type model sample (Figure 4a) is in good agreement with the report of Woodcock et al. (2007) [19]. On the other hand, Zn was floated together with Cu in the flotation of the $PbSO_4$-type model sample (Figure 4b), indicating that the floatability of sphalerite increased in the presence of anglesite, and this increased floatability of sphalerite was also observed in the real sample (sample A, Figure 3). In the case of the $PbSO_4$-type model sample (Figure 4b), the recovery of Pb was lower than 20% because anglesite has low affinity with xanthate [29,30]. Trahar et al. (1997) [23] and Wills and Napier–Munn (2005) [31] reported that the floatability of sphalerite can be increased by metal ions (e.g., Cu^{2+} and Pb^{2+}) due to the formation of CuS/PbS-like compounds which have higher affinity with xanthate than ZnS. In the case of the $PbSO_4$-type model sample, sphalerite may be activated by Pb^{2+} released from anglesite because of its higher solubility than that of PbS (Ksp of anglesite and galena are $10^{-7.79}$ and $10^{-26.77}$, respectively) [32]. These results indicate that the conventional flotation procedure of Cu-Pb-Zn ores is not applicable for SMS ores and needs to be modified to minimize the effect of anglesite on the floatability of sphalerite.

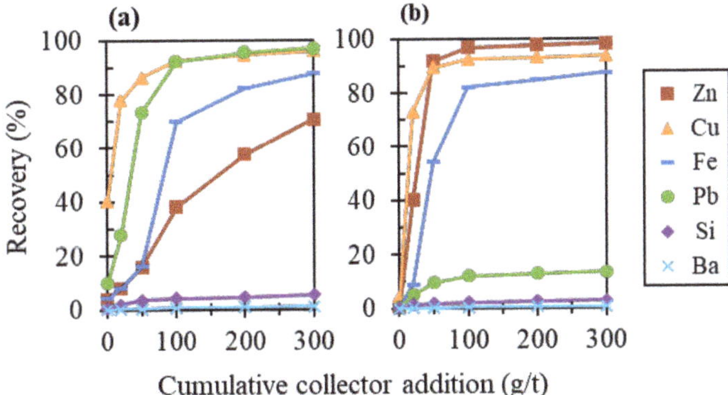

Figure 4. Flotation results of (**a**) PbS-type model sample and (**b**) $PbSO_4$-type model sample.

3.2. Leachability Test of Sample A with DI Water

Leachability test of sample A with DI water was conducted to confirm how much dissolved metal ions are released from sample A. As shown in Table 3, the concentrations of Pb^{2+} and Zn^{2+} were 38 ppm and 20 ppm, respectively, while the concentrations of Cu^{2+} and $Fe^{2+/3+}$ were below the detection limit of ICP-AES (final pH: 5.12). Fuchida et al. (2018) [11] reported that Pb^{2+} and Zn^{2+} were released in the saline water from sulfide samples collected by seafloor drilling from the Izena Hole in the middle Okinawa Trough, Japan. Other authors also reported the release of metal ions like Cu^{2+}, Pb^{2+} and Zn^{2+} from the samples obtained from SMS deposits such as the Trans-Atlantic Geotraverse (TAG) active mound on the Mid-Atlantic Ridge [10] and Hakurei Site in the Okinawa Trough, Japan [16]. These suggest that metal ions are released from SMS ores, which supports our deduction that Pb^{2+} and Zn^{2+} are released from sample A. According to Aikawa et al. (2020) [33], Rashchi et al. (2002) [22], and Trahar et al. (1997) [23], lead activation of sphalerite can occur above 5 ppm of Pb^{2+}, resulting in dramatic increase in the floatability of sphalerite. This indicates that activation of sphalerite by Pb^{2+} would occur during flotation of sample A, making not only $CuFeS_2$ but also ZnS float, so their separation becomes difficult in the presence of soluble Pb minerals like anglesite.

Table 3. Results of leaching tests of sample A with DI water.

Concentration (ppm)				Final pH
Cu	Zn	Pb	Fe	
-	20	38	-	5.12

Note: "-" denotes below the detection limit.

3.3. Effect of Surface Cleaning Pretreatment Using EDTA on the Separation of Chalcopyrite and Sphalerite in the Flotation of Sample A

The SMS ores have undergone the natural oxidation process under atmosphere and/or seafloor, resulting in the formation of oxidized phases [10,34]. According to Fuchida et al. (2019) [35] who compared metal leaching of both non-oxidized (non-exposed to atmosphere before and during exploitation) and oxidized (exposed to atmosphere after lifting and recovery) seafloor hydrothermal sulfides, the oxidized sulfides readily released large amounts of various metal(loid)s (e.g., Mn, Fe, Zn, Cu, As, Sb, and Pb) compared to non-oxidized ones. These suggest that oxidation products (e.g., oxide, hydroxide, sulfate, and carbonate) may be deposited on the surface of minerals and/or contained in sample A. These oxidation products present on mineral surface would contribute to the decrease in the floatability of sulfide minerals, especially chalcopyrite [36–39]. In addition, sample A contains the oxidation product, anglesite (lead sulfate), which makes the separation of chalcopyrite and sphalerite difficult due to the improved floatability of the latter by lead activation. To minimize the effects of oxidation products, surface cleaning pretreatment using EDTA was applied prior to flotation tests with the aim of improving the floatability of chalcopyrite, as well as depressing the floatability of sphalerite. EDTA was used for surface cleaning because of its ability to form stable complexes with metal ions dissolved from anglesite (lead sulfate)—a problematic mineral for Cu-Zn separation—as well as other oxidation products, while it does not react with metal sulfides [37,40–42].

Figure 5 shows the flotation results of sample A with and without surface cleaning pretreatment using EDTA. The dosage of sodium sulfite as a depressant in this flotation experiment was fixed at 20 kg/t. After EDTA washing, chalcopyrite was floated first followed by sphalerite (Figure 5b). Moreover, the recovery of Pb minerals was high (i.e., ~80% at 100 g/t KAX) compared to that without EDTA washing (Figure 5a). This increase in the floatability of Pb minerals is due most likely to the dissolution of most of anglesite after EDTA washing. Comparing the XRD patterns of sample A before and after EDTA washing (Figure 6), the latter showed that the peak intensity of anglesite apparently decreased while that of galena was not changed. This implies that the ratio of galena/anglesite increased, so the recovery of Pb minerals became high. Not only Pb minerals, but also the recovery of chalcopyrite at 20 g/t KAX increased from 19% to 81% after EDTA washing (Figure 5a,b). This may have been achieved by the removal of oxidation products present on the surface of chalcopyrite by EDTA. After EDTA washing, leachate contains a large amount of dissolved Pb (3200 ppm) with minor amounts of other metals (e.g., [Cu^{2+}], 9 ppm; [Zn^{2+}], 137 ppm; [$Fe^{2+/3+}$], 6 ppm) (Table 4), confirming that oxidation products and anglesite were dissolved after EDTA washing (Figure 6).

Due to the increase in the floatability of chalcopyrite, the separation efficiency of chalcopyrite and sphalerite was improved; however, the depressive effect of EDTA washing on the floatability of sphalerite was limited. In the flotation of sample A with EDTA washing, Zn was recovered as froth together with Pb, suggesting that sphalerite may be activated by Pb^{2+} forming PbS-like compounds on the surface of sphalerite. To further improve the separation efficiency of chalcopyrite and sphalerite, the depression of lead-activated sphalerite by zinc sulfate, a common depressant for sphalerite, was investigated in the next subsection.

Table 4. The concentrations of dissolved metals in the supernatant after surface cleaning with EDTA.

Concentration (ppm)			
Cu	Zn	Pb	Fe
9	137	3200	6

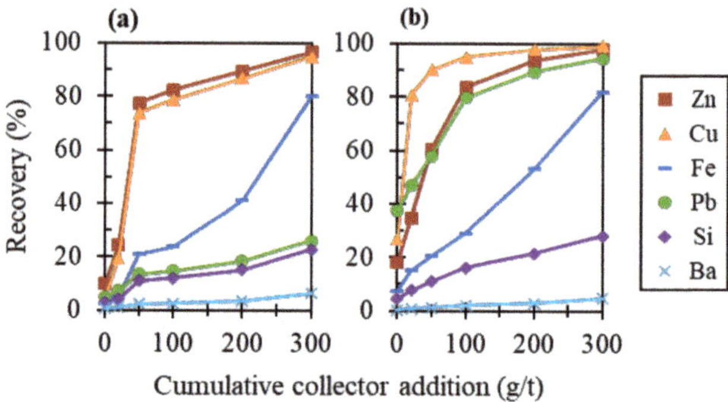

Figure 5. Flotation results of sample A (**a**) without surface cleaning and (**b**) with ethylene diamine tetra acetic acid (EDTA) washing.

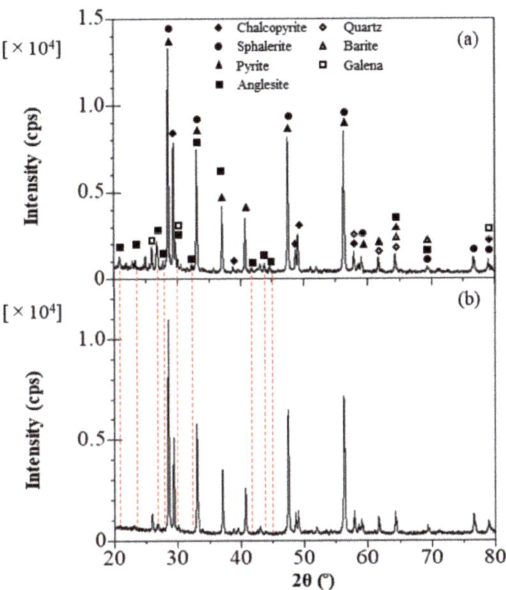

Figure 6. XRD patterns of sample A (**a**) without and (**b**) with EDTA washing.

3.4. Suppression of Lead-Activated Sphalerite by Zinc Sulfate after EDTA Washing

Figure 7a–c show the flotation results of sample A without both EDTA washing and the addition of zinc sulfate (a), with the addition of zinc sulfate (1000 ppm of Zn^{2+}) (b), and with EDTA washing followed by the addition of zinc sulfate (1000 ppm of Zn^{2+}) (c). As illustrated in Figure 7a,b, the effect of zinc sulfate in depressing the floatability of sphalerite was almost negligible; that is, sphalerite was recovered as froth together with chalcopyrite. However, when EDTA washing was adopted prior to the addition of zinc sulfate, the floatability of chalcopyrite was not affected, while it had a detrimental effect on the floatability of sphalerite (Figure 7c); for example, the recovery of sphalerite at 100 g/t KAX decreased from 88% to 8% by employing EDTA washing and the addition of zinc sulfate (Figure 7b,c). These results indicate that EDTA washing followed by Zn^{2+} addition could be an effective approach to depress the floatability of lead-activated sphalerite in the flotation of SMS ores which contain anglesite.

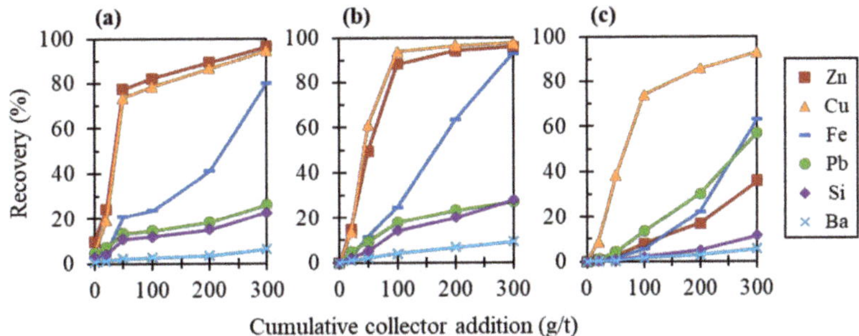

Figure 7. Flotation results of sample A (**a**) without both EDTA washing and the addition of zinc sulfate, (**b**) with the addition of zinc sulfate, and (**c**) with EDTA washing followed by the addition of zinc sulfate. Note that the recoveries of all minerals at 0 g/t of collector in Figure 7b,c were assumed to be zero due to the lack of froth amounts for the XRF analysis.

El–Shall et al. (2000) [25] calculated the change in free energy based on the equilibrium constant (Equations (1) and (2), where K = 1000) to evaluate the possibility of depression of lead-activated sphalerite in the presence of 1×10^{-4} mol/L of Pb^{2+} and 1×10^{-4} mol/L of Zn^{2+} at about pH 7.0. It was concluded that depression of lead-activated sphalerite by zinc sulfate is most probably due to the following reaction: $PbS_{surface} + Zn^{2+} \rightarrow ZnS + Pb^{2+}$. Basilio et al. (1996) [43] also calculated the equilibrium constant of Equation (1) to evaluate the possibility of lead activation of sphalerite based on the change in free energy (Equation (2)); however, the calculated value of K was different from the one used by El–Shall et al. (2000) [25]. As confirmed by the above two cases, the free energy strongly depends on the conditions (e.g., pH and concentrations of reactants and products), so the change in free energy in our flotation system needs to be calculated using the measured values in this study—4.08 ppm of Pb^{2+} and 1070 ppm of Zn^{2+} obtained from the suspension after addition of zinc sulfate in the flotation without EDTA washing (Figure 7b), and 1.21 ppm of Pb^{2+} and 949 ppm of Zn^{2+} obtained from that with EDTA washing (Figure 7c). For calculating the change in free energy in the flotation system of this study, the following equilibrium constants were considered: 1000 [25], 704 [44], 1059 [45], and 1127 [46]—the last three values were calculated based on Equation (3) using K_{sp} values of ZnS and PbS summarized in Table 5 [43–46].

$$\Delta G = -RT \ln K + RT \ln(Zn^{2+}/Pb^{2+}) \tag{2}$$

$$K = \frac{K_{sp}^{ZnS}}{K_{sp}^{PbS}} \tag{3}$$

Table 5. Ksp values of ZnS and PbS in the literature.

K_{sp}^{ZnS}	K_{sp}^{PbS}	Reference
1.9×10^{-26}	2.7×10^{-29}	Helgeson (1969) [44]
7.2×10^{-26}	6.8×10^{-29}	Latimer (1952) [45]
7.1×10^{-26}	6.3×10^{-29}	Leckie & James (1974) [46]

As shown in Table 6, the calculated values of change in free energy in the flotation with the addition of zinc sulfate (Figure 7b) were negative except the value using the equilibrium constant reported by Helgeson (1969) [44]. On the other hand, those in the flotation with the addition of zinc sulfate after EDTA washing (Figure 7c) were all positive. These results support our flotation results that the depression of lead-activated sphalerite was only achieved by the combination of EDTA washing which decreased Pb^{2+} concentration due to the removal of anglesite and the addition of zinc sulfate due to the reverse reaction of Equation (1).

Table 6. The calculation results of the change in free energy at pH 6.5 in the flotation system of sample A.

K	ΔG (kJ/mol) With Zinc Sulfate and Without EDTA Washing (Figure 7b)	ΔG (kJ/mol) With Zinc Sulfate after EDTA Washing (Figure 7c)
1000	−0.46	2.26
704	0.41	3.13
1059	−0.60	2.11
1127	−0.76	1.96

Figure 8 shows the relationship between Cu recovery and Zn recovery in flotation experiments using EDTA washing and/or zinc sulfate addition. The efficiencies of pretreatments on the separation of Cu and Zn were in the following orders: with zinc sulfate after EDTA washing >with EDTA washing >with zinc sulfate. Thus, it can be concluded that EDTA washing improves the recovery of chalcopyrite, removes anglesite, and enables the depression of lead-activated sphalerite by the addition of zinc sulfate. This proposed flotation procedure to separate chalcopyrite and sphalerite in the presence of soluble compounds like anglesite could be applied to not only SMS ores but also Cu-Pb-Zn ores in terrestrial deposits, which contain soluble compounds formed by the natural oxidations of minerals.

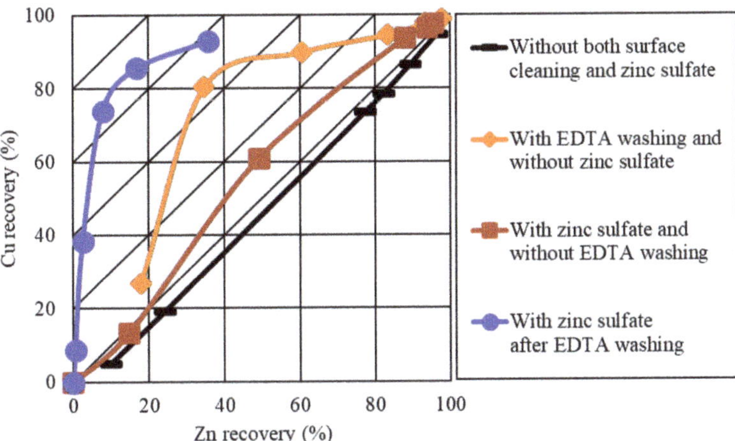

Figure 8. Relationship between Cu recovery and Zn recovery in flotation experiments with sodium sulfite 20 kg/t.

While El–Shall et al. (2000) [25] and Basilio et al. (1996) [43] estimated the possibility of lead activation of sphalerite using the measured ion concentration of Pb^{2+} and Zn^{2+}, Trahar et al. (1997) [23] reported that lead activation of sphalerite may occur even when solubility of Pb^{2+} is extremely low (e.g., at pH 10) where Pb precipitates like lead hydroxide are present. This means that sphalerite would be activated by not only Pb^{2+} but also Pb-precipitates (e.g., lead hydroxide and lead sulfate (anglesite)), and thus it is impossible to estimate whether lead activation of sphalerite occurs or not based on the concentration of dissolved Pb species. In other words, the required amount of Zn^{2+} to facilitate the depression of lead-activated sphalerite may increase when Pb-precipitates co-exist. As described above, sample A contains secondary products (e.g., oxidation products) and soluble Pb-bearing minerals (e.g., anglesite), both of which could be almost removed by EDTA washing. When the contents of soluble Pb-bearing minerals are high, however, EDTA washing cannot completely remove all the soluble Pb-bearing minerals, indicating that large amounts of residual Pb-bearing minerals most likely remain in the system. Therefore, detailed studies addressing the effects of co-existence of soluble Pb- bearing minerals on the suppression of sphalerite flotability by zinc sulfate will be of topical importance in the future.

4. Conclusions

This study investigated the applicability of surface cleaning with the addition of depressants for flotation separation of chalcopyrite and sphalerite from SMS ores. The findings of this study can be summarized as follows:

1. The obtained SMS ore sample contains $CuFeS_2$, ZnS, FeS_2, SiO_2, and $BaSO_4$ in addition to PbS and $PbSO_4$ as Pb minerals. Not only these minerals but soluble compounds which release Cu^{2+}, Zn^{2+}, Pb^{2+}, and $Fe^{2+/3+}$ are also contained in the sample.
2. When anglesite co-existed, lead activation of sphalerite occurred, which made the floatability of sphalerite increase.
3. In the flotation of sample A with sodium sulfite as a depressant for Zn- and Fe-minerals, the floatability of pyrite could be suppressed, while it was not able to depress the floatability of sphalerite because Pb^{2+} released from anglesite and other soluble compounds activated sphalerite.

4. Surface cleaning using EDTA was effective in removing anglesite and improving the recovery of chalcopyrite by dissolving secondary products formed via natural oxidation processes. However, sphalerite was floated together with chalcopyrite, even after EDTA washing.
5. The proposed flotation procedure of SMS ores, a combination of surface cleaning with EDTA to improve chalcopyrite floatability and remove anglesite and the depression of lead-activated sphalerite by using zinc sulfate, could achieve high separation efficiency of chalcopyrite and sphalerite.

Author Contributions: Conceptualization, K.A., M.I., I.P., T.O., T.T., H.F., and N.H.; methodology, K.A. and M.I.; investigation, K.A. and A.K.; data curation, K.A., A.K., and I.P.; writing—original draft preparation, K.A.; writing—review and editing, M.I. and I.P., T.O., T.T., H.F., and N.H. All authors have read and agreed to the published version of the manuscript.

Funding: This research was funded by the Agency for Natural Resources and Energy, the Ministry of Economy, Trade and Industry (METI).

Data Availability Statement: Restrictions apply to the availability of these data. Data was obtained from the Ministry of Economy, Trade and Industry (METI) and are available from the authors with the permission of METI.

Acknowledgments: This work was carried out as commissioned business by the Agency for Natural Resources and Energy, the Ministry of Economy, Trade and Industry (METI). The authors are grateful to the agency and all of the persons engaged in its duties.

Conflicts of Interest: The authors declare no conflict of interest.

References

1. Boschen, R.E.; Rowden, A.A.; Clark, M.R.; Gardner, J.P.A. Mining of deep-sea seafloor massive sulfides: A review of the deposits, their benthic communities, impacts from mining, regulatory frameworks and management strategies. *Ocean Coast. Manag.* **2013**, *84*, 54–67. [CrossRef]
2. Herzig, P.M.; Hannington, M.D. Polymetallic massive sulfides at the modern seafloor a review. *Ore Geol. Rev.* **1995**, *10*, 95–115. [CrossRef]
3. Halbach, P.; Pracejus, B.; Maerten, A. Geology and mineralogy of massive sulfide ores from the central Okinawa Trough, Japan. *Econ. Geol.* **1993**, *88*, 2210–2225. [CrossRef]
4. Janecky, D.R.; Seyfried, W.E. Formation of massive sulfide deposits on oceanic ridge crests: Incremental reaction models for mixing between hydrothermal solutions and seawater. *Geochim. Cosmochim. Acta* **1984**, *48*, 2723–2738. [CrossRef]
5. Leybourne, M.I.; de Ronde, C.E.J.; Wysoczanski, R.J.; Walker, S.L.; Timm, C.; Gibson, H.L.; Layton-Matthews, D.; Baker, E.T.; Clark, M.R.; Tontini, F.C.; et al. Geology, Hydrothermal Activity, and Sea-Floor Massive Sulfide Mineralization at the Rumble II West Mafic Caldera. *Econ. Geol.* **2012**, *107*, 1649–1668. [CrossRef]
6. German, C.R.; Petersen, S.; Hannington, M.D. Hydrothermal exploration of mid-ocean ridges: Where might the largest sulfide deposits be forming? *Chem. Geol.* **2016**, *420*, 114–126. [CrossRef]
7. Hannington, M.; Jamieson, J.; Monecke, T.; Petersen, S.; Beaulieu, S. The abundance of seafloor massive sulfide deposits. *Geology* **2011**, *39*, 1155–1158. [CrossRef]
8. Monecke, T.; Petersen, S.; Hannington, M.D. Constraints on Water Depth of Massive Sulfide Formation: Evidence from Modern Seafloor Hydrothermal Systems in Arc-Related Settings. *Econ. Geol.* **2014**, *109*, 2079–2101. [CrossRef]
9. Asakawa, E.; Murakami, F.; Tsukahara, H.; Mizohata, S. Development of vertical cable seismic (VCS) system for seafloor massive sulfide (SMS). In Proceedings of the 2014 Oceans—St. John's, St. John's, NL, Canada, 14–19 September 2014; pp. 1–7.
10. Fallon, E.K.; Niehorster, E.; Brooker, R.A.; Scott, T.B. Experimental leaching of massive sulphide from TAG active hydrothermal mound and implications for seafloor mining. *Mar. Pollut. Bull.* **2018**, *126*, 501–515. [CrossRef]
11. Fuchida, S.; Ishibashi, J.; Shimada, K.; Nozaki, T.; Kumagai, H.; Kawachi, M.; Matsushita, Y.; Koshikawa, H. Onboard experiment investigating metal leaching of fresh hydrothermal sulfide cores into seawater. *Geochem. Trans.* **2018**, *19*, 15. [CrossRef] [PubMed]
12. Zha, L.; Li, H.; Wang, N. Electrochemical Study of Galena Weathering in NaCl Solution: Kinetics and Environmental Implications. *Minerals* **2020**, *10*, 416. [CrossRef]
13. Collins, P.C.; Croot, P.; Carlsson, J.; Colaço, A.; Grehan, A.; Hyeong, K.; Kennedy, R.; Mohn, C.; Smith, S.; Yamamoto, H.; et al. A primer for the Environmental Impact Assessment of mining at seafloor massive sulfide deposits. *Mar. Policy* **2013**, *42*, 198–209. [CrossRef]
14. Narita, T.; Oshika, J.; Toyohara, T.; Okamoto, N.; Shirayama, Y. The Environmental Impact Assessment for Mining Seafloor Massive Sulphides in Japan. *J. MMIJ* **2015**, *131*, 634–638. [CrossRef]

15. Van Dover, C.L. Impacts of anthropogenic disturbances at deep-sea hydrothermal vent ecosystems: A review. *Mar. Environ. Res.* **2014**, *102*, 59–72. [CrossRef] [PubMed]

16. Ministry of Economic, Trade and Industry (METI); Japan Oil, Gas and Metals National Corporation (JOGMEC). Summary Report on Comprehensive Evaluations of Development Plan of Seafloor Massive Sulfide Deposits. Available online: http://www.jogmec.go.jp/content/300359550.pdf (accessed on 26 January 2021).

17. Oki, T.; Nishisu, Y.; Hoshino, M. Characteristics of The Existing Mineral Phases of Japanese Submarine Hydrothermal Polymetallic Sulfides and Their Influence on Respective Mineral Processing Properties. *J. MMIJ* **2015**, *131*, 619–626. [CrossRef]

18. Masuda, N. Challenges toward the sea-floor massive sulfide mining with more advanced technologies. In Proceedings of the 2011 IEEE Symposium on Underwater Technology and Workshop on Scientific Use of Submarine Cables and Related Technologies, Tokyo, Japan, 5–8 April 2011; pp. 1–4.

19. Woodcock, J.T.; Sparrow, G.J.; Bruckard, W.J.; Johnson, N.W.; Dunne, R. Plant Practice: Sulfide Minerals and Precious Metals. In *Froth Flotation a Century of Innovation*; Fuerstenau, M., Jameson, G., Yoon, R.H., Eds.; SME: Littleton, CO, USA, 2007; pp. 781–843. ISBN 978-0-873-35252-9.

20. Houot, R.; Raveneau, P. Activation of sphalerite flotation in the presence of lead ions. *Int. J. Miner. Process.* **1992**, *35*, 253–271. [CrossRef]

21. Laskowski, J.S.; Liu, Q.; Zhan, Y. Sphalerite activation: Flotation and electrokinetic studies. *Miner. Eng.* **1997**, *10*, 787–802. [CrossRef]

22. Rashchi, F.; Sui, C.; Finch, J.A. Sphalerite activation and surface Pb ion concentration. *Int. J. Miner. Process.* **2002**, *67*, 43–58. [CrossRef]

23. Trahar, W.J.; Senior, G.D.; Heyes, G.W.; Creed, M.D. The activation of sphalerite by lead—A flotation perspective. *Int. J. Miner. Process.* **1997**, *49*, 121–148. [CrossRef]

24. Wang, H.; Wen, S.; Han, G.; Xu, L.; Feng, Q. Activation mechanism of lead ions in the flotation of sphalerite depressed with zinc sulfate. *Miner. Eng.* **2020**, *146*, 106132. [CrossRef]

25. El-Shall, H.E.; Elgillani, D.A.; Abdel-Khalek, N.A. Role of zinc sulfate in depression of lead-activated sphalerite. *Int. J. Miner. Process.* **2000**, *58*, 67–75. [CrossRef]

26. Fuerstenau, D.W.; Metzger, P.H.; Fuerstenau, D.W.; Metzger, P.H. Activation of sphalerite with lead ions in the presence of zinc salts. *Trans. Am. Inst. Min. Metall. Eng.* **1960**, *217*, 119–123.

27. Cao, Q.; Cheng, J.; Feng, Q.; Wen, S.; Luo, B. Surface cleaning and oxidative effects of ultrasonication on the flotation of oxidized pyrite. *Powder Technol.* **2017**, *311*, 390–397. [CrossRef]

28. Clarke, P.; Fornasiero, D.; Ralston, J.; Smart, R.S.C. A study of the removal of oxidation products from sulfide mineral surfaces. *Miner. Eng.* **1995**, *8*, 1347–1357. [CrossRef]

29. Fuerstenau, M.C.; Olivas, S.A.; Herrera-Urbina, R.; Han, K.N. The surface characteristics and flotation behavior of anglesite and cerussite. *Int. J. Miner. Process.* **1987**, *20*, 73–85. [CrossRef]

30. Rashchi, F.; Dashti, A.; Arabpour-Yazdi, M.; Abdizadeh, H. Anglesite flotation: A study for lead recovery from zinc leach residue. *Miner. Eng.* **2005**, *18*, 205–212. [CrossRef]

31. Wills, B.A.; Napier-Munn, T.J. Froth flotation. In *Mineral Processing Technology*, 8th ed.; Elsevier: Oxford, UK, 2016; pp. 265–380. ISBN 978-0-08-097053-0.

32. Ball, J.W.; Nordstrom, D.K. *User's Manual for WATEQ4F, with Revised Thermodynamic Data Base and Text Cases for Calculating Speciation of Major, Trace, and Redox Elements in Natural Waters*; USGS Numbered Series Open-File Report 91-183; U.S. Geological Survey: Reston, VA, USA, 1991; p. 37. [CrossRef]

33. Aikawa, K.; Ito, M.; Segawa, T.; Jeon, S.; Park, I.; Tabelin, C.B.; Hiroyoshi, N. Depression of lead-activated sphalerite by pyrite via galvanic interactions: Implications to the selective flotation of complex sulfide ores. *Miner. Eng.* **2020**, *152*, 106367. [CrossRef]

34. Seal, R.R.; Foley, N.K. *Progress on Geoenvironmental Models for Selected Mineral Deposit Types*; USGS Numbered Series Open-File Report 02-195; U.S. Geological Survey: Reston, VA, USA, 2002; pp. 196–212. [CrossRef]

35. Fuchida, S.; Ishibashi, J.; Nozaki, T.; Matsushita, Y.; Kawachi, M.; Koshikawa, H. Metal Mobility from Hydrothermal Sulfides into Seawater During Deep Seafloor Mining Operations. In *Environmental Issues of Deep-Sea Mining*; Sharma, R., Ed.; Springer: Cham, Switzerland, 2019; pp. 213–229. ISBN 978-3-030-12695-7.

36. Senior, G.D.; Trahar, W.J. The influence of metal hydroxides and collector on the flotation of chalcopyrite. *Int. J. Miner. Process.* **1991**, *33*, 321–341. [CrossRef]

37. Kant, C.; Rao, S.R.; Finch, J.A. Distribution of surface metal ions among the products of chalcopyrite flotation. *Miner. Eng.* **1994**, *7*, 905–916. [CrossRef]

38. Park, I.; Hong, S.; Jeon, S.; Ito, M.; Hiroyoshi, N. A Review of Recent Advances in Depression Techniques for Flotation Separation of Cu–Mo Sulfides in Porphyry Copper Deposits. *Metals* **2020**, *10*, 1269. [CrossRef]

39. Park, I.; Hong, S.; Jeon, S.; Ito, M.; Hiroyoshi, N. Flotation Separation of Chalcopyrite and Molybdenite Assisted by Microencapsulation Using Ferrous and Phosphate Ions: Part I. Selective Coating Formation. *Metals* **2020**, *10*, 1667. [CrossRef]

40. Bicak, O. A technique to determine ore variability in a sulphide ore. *Miner. Eng.* **2019**, *142*, 105927. [CrossRef]

41. Grano, S.R.; Ralston, J.; Johnson, N.W. Characterization and treatment of heavy medium slimes in the Mt. Isa mines lead-zinc concentrator. *Miner. Eng.* **1988**, *1*, 137–150. [CrossRef]

42. Rumball, J.A.; Richmond, G.D. Measurement of oxidation in a base metal flotation circuit by selective leaching with EDTA. *Int. J. Miner. Process.* **1996**, *48*, 1–20. [CrossRef]
43. Basilio, C.; Kartio, I.; Yoon, R.-H. Lead activation of sphalerite during galena flotation. *Miner. Eng.* **1996**, *9*, 869–879. [CrossRef]
44. Helgeson, H.C. Thermodynamics of hydrothermal systems at elevated temperatures and pressures. *Am. J. Sci.* **1969**, *267*, 729–804. [CrossRef]
45. Latimer, W.M. *The Oxidation States of the Elements and Their Potentials in Aqueous Solutions*, 2nd ed.; Prentice-Hall, Inc.: Upper Saddle River, NJ, USA, 1952; pp. 72, 152, 169. ISBN 978-0-758-15876-5.
46. Leckie, J.O.; James, R.O. *Aqueous Environmental Chemistry of Metals*; Rubin, A.J., Ed.; Ann Arbor Science Publishers: Ann Arbor, MI, USA, 1974; pp. 1–76. ISBN 978-0-250-40060-7.

Article

Fine Bauxite Recovery Using a Plate-Packed Flotation Column

Pengyu Zhang [1,2], Saizhen Jin [1,2], Leming Ou [1,2,*], Wencai Zhang [3,*] and Yuteng Zhu [1,2]

[1] School of Minerals Processing and Bioengineering, Central South University, Changsha 410083, China; pengyu7765@csu.edu.cn (P.Z.); jinsaizhen@csu.edu.cn (S.J.); zhuyuteng@csu.edu.cn (Y.Z.)
[2] Key Laboratory of Hunan Province for Clean and Efficient Utilization of Strategic Calcium-Containing Mineral Resources, Central South University, Changsha 410083, China
[3] Department of Mining and Minerals Engineering, Virginia Polytechnic Institute and State University, Blacksburg, VA 24061, USA
* Correspondence: olmpaper@csu.edu.cn (L.O.); wencaizhang@vt.edu (W.Z.); Tel.: +86-0731-8883-0913 (L.O.); +540-231-6671 (W.Z.)

Received: 2 August 2020; Accepted: 24 August 2020; Published: 2 September 2020

Abstract: In this investigation, the fine-grained bauxite ore flotation was conducted in a plate-packed flotation column. This paper evaluated the effects of packing-plates on recovering fine bauxite particles and revealed the fundamental mechanisms. Bubble coalescence and break-up behaviors in the packed and unpacked flotation columns were characterized by combining Computational Fluid Dynamics (CFD) and Population Balance Model (PBM) techniques. Flotation experiments showed that packing-plates in the collection zone of a column can improve bauxite flotation performance and increase the smaller bauxite particles recovery. Using packing-plates, the recovery of Al_2O_3 increased by 2.11%, and the grade of Al_2O_3 increased by 1.85%. The fraction of -20 μm mineral particles in concentrate increased from 47.31% to 54.79%. CFD simulation results indicated that the packing-plates optimized the bubble distribution characteristics and increased the proportion of microbubbles in the flotation column, which contributed to improving the capture probability of fine bauxite particles.

Keywords: plate-packed flotation column; fine bauxite particles; bubble characteristics

1. Introduction

Most of the valuable minerals in ores are required concentration to reduce the operation costs of downstream extractive metallurgical processes. The concentration process is known as the key step of mineral processing. In the field of mineral processing, recovering fine minerals is difficult thus resulting in a lot of economic losses [1,2]. To solve these problems, column flotation technology has been developed in the past few decades [3,4]. Due to the structural features, flotation columns usually provide a higher recovery of fine minerals compared with other flotation equipment [5]. The mechanism of a traditional countercurrent flotation column is to make bubbles and mineral particles move towards each other, and then the hydrophobic particles are captured by the bubbles from the slurry, and rise with the bubbles to a froth zone. With the continuous flow of bubbles, these captured mineral particles gradually get out of the column. Consequently, the bubble size and distribution characteristics in flotation column have significant effects on the concentration performance.

After decades of progress, many different types of flotation columns have been derived and used in the industrial processing of various minerals [6–8]. Flotation columns packed with plates have specific internal structures, which impose direct effects on the flow characteristics of slurry inside the column, thereby affecting the flotation process of minerals [9–11]. For instance, a honeycomb tube is designed as packing-plates in previous research works [12,13]. Experimental results showed that honeycomb tube can promote the generation of smaller bubbles, thereby increasing gas content, as well as particle–bubble collision probability. Moreover, the honeycomb tube packing causes turbulent rotating flow into a mild flow in the column, forming a static hydrodynamic environment and reducing the probability of detachment. The bubble behaviors was also investigated in a lab-scale cyclonic-static micro-bubble flotation column packed by sieve plates [14]. Based on Particle Image Velocimetry (PIV) and Charge-Coupled Device (CCD) camera techniques, it was found that packing non-uniform sieve plates was more effective in terms of bubble distribution equalization, air column inhabitation, and non-axial velocity decreasing.

In a previous study of fine bauxite ore column flotation [15], square packed-plates were installed in the collection zone of a flotation column in a multi-layer packing manner. Based on flotation experiments, the addition of packing-plates led to a better separation of bauxite from gangue minerals and a higher recovery of fine mineral particles. Computational Fluid Dynamics (CFD) simulation results indicated that the original intensively turbulent environment was weakened and dispersed into several units with different turbulent intensities in axial, which enhanced the separation of mineral particles with different properties. But the mechanism of improved recovery of fine mineral particles by the packing-plates has not been fully described and presented. In this present work, CFD combined with Population Balance Model (PBM) simulations were performed to characterize the bubble coalescence and break-up behaviors in the unpacked flotation column (UFC) and the plate-packed flotation column (PFC). Additionally, the effects of bubble diameter on recovering finer bauxite particles was also evaluated.

2. Materials and Methods

2.1. Flotation Apparatus and Procedures

Flotation experiments were conducted using a squared-plates column flotation apparatus (Figure 1). The flotation column has a dimension of $\varphi 80$ mm $\times$ 2000 mm. This apparatus primarily consists of three subsystems: a Plexiglas flotation column, an air injection system, and a slurry circulation system. The packing-plates was made by polyethylene plates (thickness δ = 1 mm); its structure was shown in different viewpoints (see Figure 1). Three layers of packing-plates were evenly installed along the column between the slurry inlet and air inlet.

Experimental procedures of the column flotation tests were described as follows:

(1) Grind 2 kg of the bauxite ore (collected from a bauxite mine located in Henan Province, Luoyang, China) until its average particle size (D_{50}) reaches about 20 μm;

(2) Condition the feed slurry in the slurry mixing tank at room temperature and pH 9.5 (sodium carbonate as pH modifier). Add 100 g/t hexametaphosphate into the feed slurry then stir for 3 min. Followed by adding 1200 g/t sodium oleate and conditioning for 4 min. All three reagents are analytical grade supplied by Aladdin Biochem. Tech., Shanghai, China. The initial slurry solid concentration is 15%;

(3) Turn on the air compressor (0.6 MPa), and adjust air inlet flowrate to 2.5 L/min; Turn on the peristaltic pump, and adjust feed flowrate to 3.34 L/min. When foams start to flow out from the top of the column, collect the froth product as concentrate K for a period of 12 min, after which the slurry remaining in the column is collected in mixing tank as tailings X. The column flotation test is running at batch mode;

(4) Filter and dry the concentrate K and tailings X, followed by elemental analysis using X-ray fluorescence. Additionally, perform particle size analysis on the concentrate K using a laser particle size analyzer (Mastersize 2000). Both instruments are from Malver Panalytical, Malvern, UK.

Additionally, bubbles were observed in an area circled in the red box shown in Figure 1. Due to the problem of poor light transmittance during the flotation process of the bauxite ore, it was hard to directly observe the distribution characteristics of bubbles. Therefore, bubbles were only observed in the tests without adding any mineral particles. Photos of the bubbles were captured using a Canon camera (EOS 800D, Canon, Tokyo, Japan).

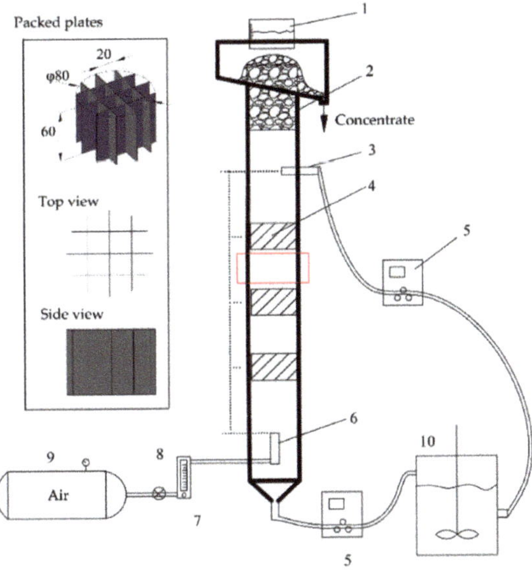

Figure 1. Experiment apparatus used for fine bauxite ore flotation: 1—Washing water device; 2—Plexiglass column; 3—Slurry inlet; 4—Packed plates; 5—Peristaltic pump; 6—Air-bubble sparger; 7—Air flowmeter; 8—Regulating valve; 9—Air compressor; 10—Slurry mixing tank.

The bauxite ore used in this research contains 44.67% of Al_2O_3 and 20.68% of SiO_2, corresponding to an aluminum oxide to silicon oxide grade (Al_2O_3/SiO_2) ratio of 2.16. Due to the relatively high content of SiO_2, the ore needs to be concentrated to improve the Al/Si ratio in order to meet the requirements on feed grade of the Bayer process. Other major components of the ore included Fe (5.19%), TiO_2 (3.76%), CaO (2.77%), K_2O (2.76%), and S (0.97%). Mica, siderite, kaolinite, anatase, quartz, calcite, and pyrite were the dominant gangue minerals.

2.2. Simulation Procedures

The simulation work was conducted in water-air two phase system using ANSYS Fluent 18.2 software (Ansys Inc., Canonsburg, PA, USA). Detailed procedures were as follows: (a) pre-processing uses ICEM for model extraction and meshing; (b) numerical calculation uses Fluent; and (c) post-processing uses CFD-post for data processing and output. This research focus on bubble characteristics affected by packing-plates in the flotation column. The collection zone of flotation column was selected as the computing domain. A geometric model of the flotation column used in the experimental apparatus was set-up as shown in Figure 2. Due to symmetry of the column in axial direction, simulation was only performed on half of the column. Additionally, boundary conditions of the calculation were based on flotation parameters. The mesh dependent tests for CFD simulation of flow calculation were conducted. The test results showed that 380,000 grids are an appropriate number to be used for UFC. Thus, a total of 380,000 grids were used for simulating flow in UFC. In case of PFC, the grid surrounding the plates was refined using the function of density box. Eulerian-Eulerian multiphase model and standard k-epsilon model were used for calculation. The bubble coalescence and break-up processes were simulated by population balance model (PBM). Parameter settings of the PBM are given in Table 1.

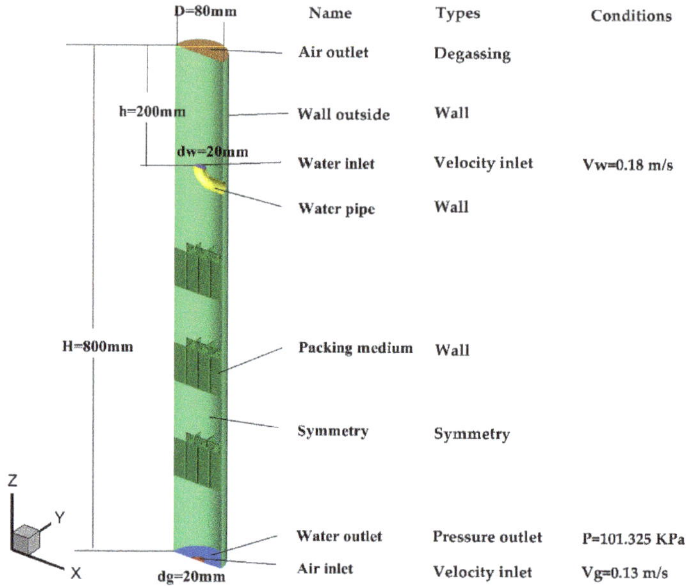

Figure 2. Geometry and boundary conditions of the flotation column for simulation.

Table 1. The parameters of population balance model (PBM).

Parameters	Value
Method	Discrete
Kv	0.52
Bubble Diameter Range	0.40~4.03 mm
Initial Bubble Diameter	1 mm
Aggregation Kernel	Luo model [16]
Breakage Kernel	Frequency-Luo model [16]; Formulation-Hagesather
Surface Tension	0.04 N/m

3. Results

3.1. Flotation Results

Flotation experiments were performed using both UFC and PFC to evaluate the effects of packing-plates on bauxite flotation performance. Figure 3 presents the flotation grade and recovery of Al_2O_3, as well as Al_2O_3/SiO_2 ratio, obtained by using UFC and PFC. It can be seen that a concentrate containing 58.4% Al_2O_3 was obtained using PFC at a recovery of 52.68%. Compared UFC, Al_2O_3 recovery and Al_2O_3 grade were increased by 2.11% and 1.85%, respectively, by using PFC. These results indicated that flotation performance of the bauxite ore was improved by packing-plates. More aluminum-enriched particles were recovered using PFC. Al_2O_3/SiO_2 ratio of the concentrated obtained using PFC reached 9.72, which is much higher than UFC. Therefore, it could be inferred that the packing-plates was able to improve the separation of aluminum minerals from silicon minerals occurring in the bauxite ore.

Particle size analysis was conducted on the concentrate obtained by using UFC and PFC. As shown in Figure 4, the fraction of −20 μm particles in UFC concentrate was 47.31 (volumetric fraction, %), however for PFC concentrate, the fraction increased up to 54.79. For particles of 20~37 μm size range, the volumetric fraction was increased by 3.06 by using PFC. Totally, the fraction of −37 μm particles in the concentrate generated from PFC increased by 10.54. Moreover, the flotation experimental results showed that PFC concentrate had a higher Al_2O_3 grade with improved recovery (see Figure 3). Based on these findings, it can be concluded that the preferential recovery of Al_2O_3 enriched particles is attributable to the increased fraction of finer bauxite particles in the flotation concentrate. The packing-plates inside the flotation column led to improvements in the recovery of fine bauxite particles.

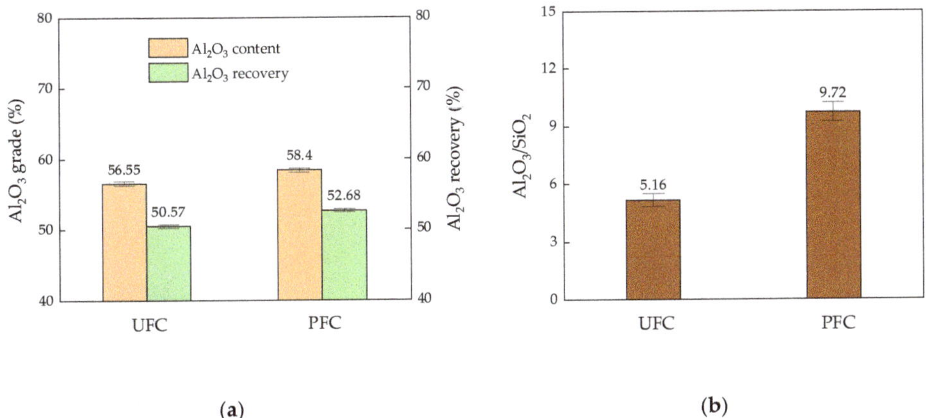

Figure 3. Flotation concentrate obtained by using unpacked flotation column (UFC) and plate-packed flotation column (PFC): (**a**) Al_2O_3 grade and recovery; (**b**) Al_2O_3/SiO_2 ratio.

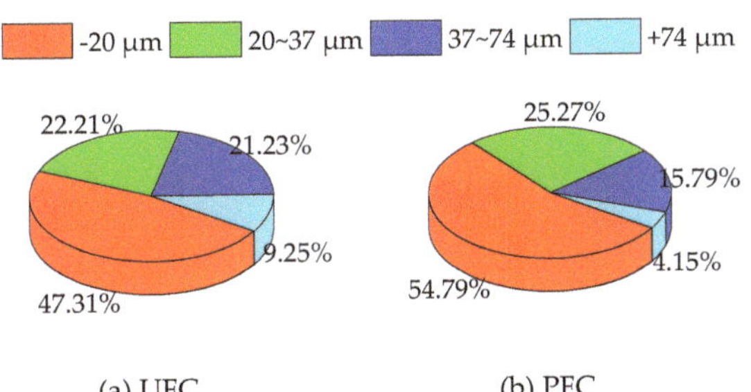

Figure 4. Particle size distribution of flotation concentrate by using (**a**) UFC and (**b**) PFC.

3.2. Bubble Characteristics

Figure 5a,b present the bubble size distribution of flows in UFC and PFC, respectively. It can be seen that bubble size increased along the axial direction and decreased along the radial direction. The bubble sizes were distributed in the range of around 0~4.2mm. When the packing-plates was used, the bubble diameter decreased in the entire flotation zone of the column. Comparing the distribution of bubbles in UFC and PFC, it can be seen that in the presence of packing-plates, bubbles in PFC have more obvious size differences, and the proportion of bubbles with smaller diameters increased. According to the simulation results, it was calculated that the mean diameter of bubbles was reduced from 3.03 mm (in UFC) to 2.80 mm (in PFC). The fractional distribution of bubbles in the column were also studied based on the simulation data. The results are shown in Figure 5c,d. It can be seen that the bubble diameter in UFC was primarily concentrated in the range of 2.5~4 mm. Bubbles with a diameter of 3.5 mm or larger accounted for the largest proportion, reaching 37.17%. When the column used packing-plates, the proportion of small-sized bubbles increased significantly, and the bubbles were primarily concentrated in the size range of 2.3~3.7 mm. It indicates that packing-plates could increase the proportion of microbubbles and reduce the average diameter of bubbles.

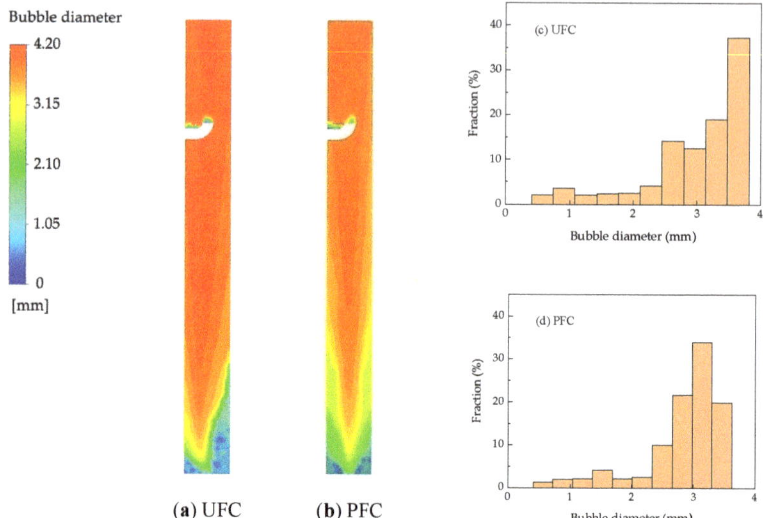

Figure 5. Bubble distribution in UFC and PFC: (**a**,**b**) Bubble size distribution in specific plane $Y = 2$ mm; (**c**,**d**) the fractional distribution of bubble size.

In order to verify the simulation results of bubble distribution in UFC and PFC, images of the bubbles generated in the columns under conditions similar to those used for the simulation were captured. Flows in the flotation column are three dimensional, however the images obtained using a digital camera are two dimensional. Therefore, only qualitative analysis was completed in this study. Based on the images shown in Figure 6, it can be observed that the fraction of large-size bubbles in UFC was less than PFC. Moreover, the coalescence between bubbles was reduced after packed-plates were used. This is consistent with the numerical simulation results. It is inferred that packing-plates in flotation column will inhibit bubble coalescence and promote the formation of small-size bubbles.

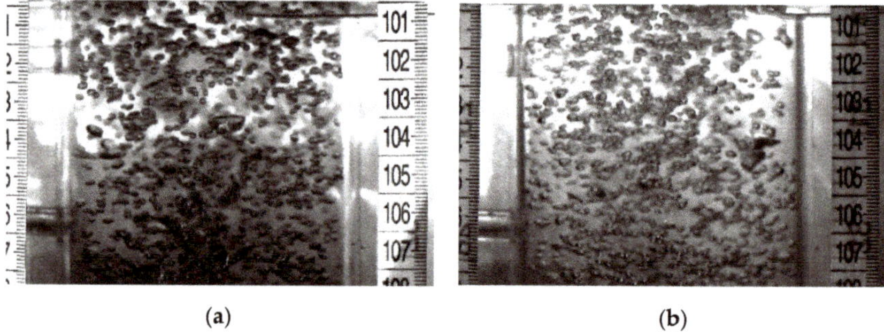

Figure 6. Images of the bubbles captured in UFC and PFC: (**a**) Bubbles in UFC; (**b**) Bubbles in PFC.

3.3. Turbulence Characteristics

The movement and collision of bubbles and particles are mainly caused by fluid pulsation. In fluid mechanics, turbulent kinetic energy k is used to characterize the pulsation of large-scale vortex. The larger the value of k, the higher the pulsation velocity of large-scale vortex. Figure 7 presents the turbulent kinetic energy (TKE) distribution characteristics in UFC and PFC. It can be seen from the figure that without packing, the areas with high values of TKE were concentrated near the central axis, and decreased along both the axial and the radial direction. TKE was higher than 1×10^{-2} m^2/s^2.

After adding the packing-plates to the flotation column, the TKE inside the entire flotation column was significantly reduced. The area associated with high-value TKE was obviously reduced, and the TKE in the plates-filled area was reduced to 3.75×10^{-3} m^2/s^2. This showed that the packed plates could effectively reduce the TKE in the collection area of the flotation column, reduce the pulsating velocity of the turbulent vortex, and form a relatively mild turbulent environment.

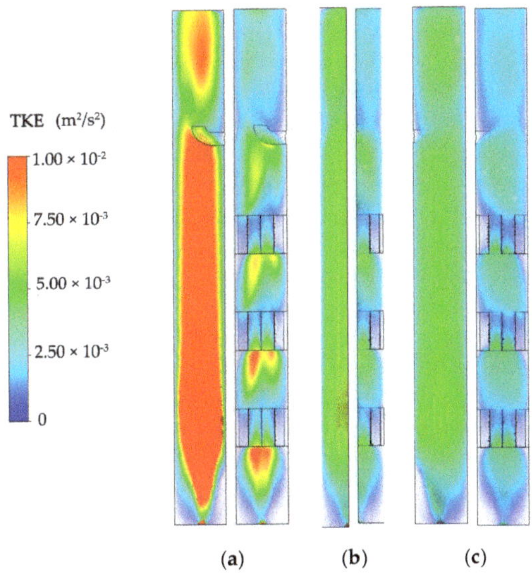

Figure 7. Turbulent kinetic energy (TKE) distribution characteristics in UFC and PFC: (**a**) Front view; (**b**) Sside view; (**c**) Rear view.

According to the simulation results, the cumulative curves of the volumetric percent of different turbulent kinetic energies are shown in Figure 8. From the results, it can be seen that the TKE in PFC was evenly distributed in the majority of the flotation column collection area. Rapid rises in the energy occurred at volume percent of 90, indicating that the volume of the area with high turbulence occupied about 10% of the collection area. In this area, the value of TKE was concentrated, the pulsating velocity of the fluid was greater, and the intensity of turbulence was greater. Compared with PFC, UFC had a higher cumulative TKE of about 0.03 m^2/s^2, which was about 3 times the value of the TKE of PFC. It showed that packing-plates in the flotation column could significantly reduce the TKE.

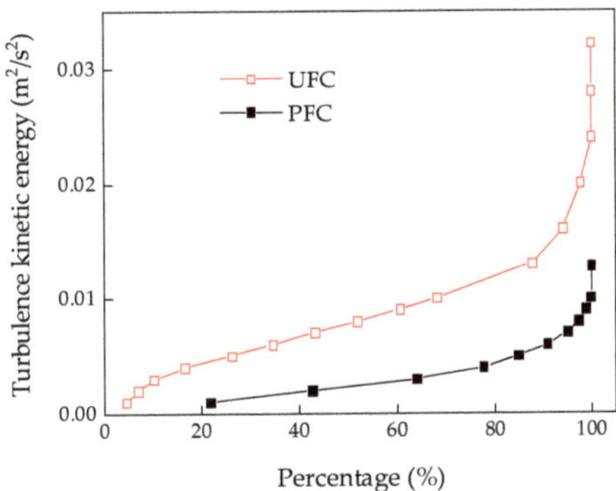

Figure 8. Turbulence kinetic energy of cumulative volume as the function of UFC and PFC.

4. Discussion

In a multiphase flow, the coalescence and break-up behaviors of bubbles are mainly affected by the surface tension of liquid phase and the hydrodynamic environment as stated by Laari and Turunen, as well as Sattar et al. [17,18]. Based on this point, a series of mathematical models about bubble coalescence and break-up have been developed by Luo and Svendsen, as well as Lehr et al. [16,19]. The main factor leading to bubble breakage is the turbulent vortex generated by the pulsation of fluid. When the turbulent vortex carries more energy than the surface energy of newly-formed bubbles, the bubble break-up may occur, and the TKE of the vortex is converted into the surface energy of the newly formed bubble. The high-energy turbulent pulsation will aggravate the collision, merging, and fragmentation behaviors of bubbles, leading to an increase in the average geometric size of the bubbles. The addition of packing-plates in the collection area of flotation column weakens the turbulent energy of flows, thus forming a milder turbulent environment (see Figure 8). Consequently, it contributes to the formation of small-size bubbles in flotation column.

According to the reports of Zhang et al. [12,13], using packed-honeycomb tubes in flotation column can increase the gas holdup of column and generate small-size bubbles. Thereby, the packed cyclonic-static micro bubble flotation column performs better in copper sulfide flotation. Xia et al. [10] performed a two-dimensional Euler–Lagrangian model to simulate the multiphase flow for some cases of baffled and packed columns. It has been found that the presence of baffles and packing will trap bubbles or hinder the upward movement of bubbles and increase the gas hold up. Combined with the simulation and test results in this study, the average diameter of the bubble group decreased after the flotation column was packed with a plates. It is easy to infer that the number of bubbles in the collection area was increased after use of packing-plates. For mineral flotation process, a higher bubble number and a smaller bubble size usually mean that the mineral particles have a higher capture rate. For fine mineral particles, the capture rate can be calculated as

$$P = P_c P_a. \tag{1}$$

The particle collision rate formula is given by Tao et al. [20]:

$$P_c = \left(\frac{3}{2} + \frac{4\mathrm{Re}_b^{0.72}}{15}\right)\left(\frac{R_p}{R_b}\right)^2. \tag{2}$$

The particle adhesion rate formula is given by Yoon and Luttrell [21]:

$$P_a = \sin^2\left\{2\arctan\exp\left[\frac{-(45 + 8\mathrm{Re}_b^{0.72})U_b t_i}{30R_b(R_b/R_p + 1)}\right]\right\}, \tag{3}$$

where Re_b is the Reynolds number of bubble, and the formula is as follows:

$$\mathrm{Re}_b = \frac{\rho_f U_b d_b}{\mu_f}. \tag{4}$$

The relative velocity of bubble is calculated using the following formula given by Schubert, as well as Shubert and Bischofberger [22,23]:

$$\sqrt{U_b^2} = 0.33\frac{\varepsilon^{4/9} d_b^{7/9}}{\nu_f^{1/3}}\left(\frac{\rho_b - \rho_f}{\rho_f}\right)^{2/3}. \tag{5}$$

The particle induction time formula given by Koh and Schwarz [24] is

$$t_i = \frac{75}{\theta} d_p^{0.6}. \tag{6}$$

An approximate relationship exists between the particle capture rate P_{PFC} with packing and the capture rate P_{UFC} without packing. It can be expressed as

$$P_{PFC} = CP_{UFC},\tag{7}$$

where C is "Correlation coefficient". According to simulation results given above, the specific values are given as follows: $\mu_f = 1.003 \times 10^{-3}$ Pa·s; $v_f = 1.004 \times 10^{-6}$ m^2/s; $\rho_f = 998.2$ kg/m^3; $\rho_b = 1.225$ kg/m^3; and $d_p = 20$ μm; $\theta = 60°$. Turbulent dissipation rate ε and bubble diameter d_b are calculated based on the average value from the numerical simulation results. For UFC $\varepsilon_1 = 3 \times 10^{-5}$ m^2/s^3, $d_{b1} = 3.03$ mm; for PFC $\varepsilon_2 = 1 \times 10^{-5}$ m^2/s^3, $d_{b2} = 2.80$ mm. Substituting the specific values into the formula, it can be calculated $C = 1.14$. As shown in Equation (7), this means, for a mineral particle with a diameter of 20 μm, the capture probability with packing-plates is 1.14 times that of without packing.

In flotation process, reducing the size of the bubbles can not only improve the collection capacity of finer mineral particles, but also increase the volumetric concentration of bubbles in the slurry and improve the capture probability. Giving an analysis of bauxite flotation concentrate results (Figure 3) and the particle size distribution results (Figure 4), it can be seen that in PFC, the recovery of particles of smaller than 37 μm was increased by 10.54%. Among them, the proportion of particles smaller than 20 μm increased by 7.48%. The Al$_2$O$_3$ grade in the concentrate increased by 1.85%, and the recovery increased by 2.11%. These findings collectively confirmed that packing-plates inside the flotation column can increased the recovery of fine-grained mineral particles by reducing bubble diameters.

5. Conclusions

In this research, PBM incorporated with CFD techniques was performed to characterize the bubble coalescence and break-up behaviors in UFC and PFC. The effects due to application of the squared packed-plates on recovering fine bauxite particles was evaluated, and understanding of the fundamental mechanisms are achieved. The conclusions are as follows:

1. The packing-plates can significantly reduce the turbulent kinetic energy and promote the formation of a milder turbulent environment in the collection area. This will weaken the collision, merging, and fragmentation behaviors of bubbles, contributing to the formation of small-sized bubbles in flotation column.
2. Packing-plates can optimize bubble size distribution in the flotation column, and increase the proportion of micro-bubbles. According to the simulation results, the mean diameter of bubbles was reduced from 3.03 mm to 2.80 mm by packing-plates in flotation column. For mineral particles with a diameter of 20 μm, the capture probability with packing-plates is 1.14 times that of without packing.
3. Packing-plates in the collection zone of a column can improve bauxite flotation performance and enhance the recovery of fine bauxite particles. With packing, the Al$_2$O$_3$ recovery increased by 2.11%, and the Al$_2$O$_3$ grade increased by 1.85%.

Author Contributions: Conceptualization, methodology and formal analysis, P.Z.; investigation, S.J. and Y.Z.; writing—original draft preparation, P.Z.; formal analysis and writing—review and editing, W.Z.; supervision and project administration, L.O. All authors have read and agreed to the published version of the manuscript.

Funding: This work was financially supported by the National Natural Science Foundation of China (No. 51674291), and the Fundamental Research Funds for the Central Universities of Central South University (No. 2017zzts009). Key Laboratory of Hunan Province for Clean and Efficient Utilization of Strategic Calcium-containing Mineral Resources (No. 2018TP1002).

Acknowledgments: We acknowledge Beijing Lanwei Technology Co., Ltd. for providing access to ANSYS-Fluent program.

Conflicts of Interest: The authors declare no conflict of interest.

Nomenclature

φ	Diameter of column (mm)
δ	Thickness of plate(mm)
k	Turbulent kinetic energy (m^2/s^2)
P	Capture rate (%)
P_c	Collision rate (%)
P_a	Adhesion rate (%)
R_p	Particle radius (μm)
R_b	Bubble radius (mm)
Re_b	Reynolds number of bubbles
U_b	Relative velocity of bubble (m/s)
t_i	Induction time (s)
θ	Contact angle of mineral ($^\circ$)
ε	Turbulent energy dissipation rate (m^2/s^3)
ρ_f	Fluid density (Kg/m^3)
μ_f	Dynamic viscosity of fluid (Pa·s)
ν_f	Kinematic viscosity of fluid (m^2/s)
d_b	Bubble diameter (mm)
ρ_b	Bubble density (Kg/m^3)
P_{PFC}	Capture rate of PFC
P_{UFC}	Capture rate of UFC

References

1. Miettinen, T.; Ralston, J.C.; Fornasiero, D. The limits of fine particle flotation. *Miner. Eng.* **2010**, *23*, 420–437. [CrossRef]
2. Santana, R.C.; Duarte, C.R.; Ataíde, C.H.; Barrozo, M.A.S. Flotation selectivity of phosphate ore: Effects of particle size and reagent concentration. *Sep. Sci. Technol.* **2011**, *46*, 1511–1518. [CrossRef]
3. Harbort, G.; Clarke, D. Fluctuations in the popularity and usage of flotation columns—An overview. *Miner. Eng.* **2017**, *100*, 17–30. [CrossRef]
4. Prakash, R.; Majumder, S.K.; Singh, A. Flotation technique: Its mechanisms and design parameters. *Chem. Eng. Process. Process Intensif.* **2018**, *127*, 249–270. [CrossRef]
5. Yianatos, J.B. Fluid flow and kinetic modelling in flotation related processes: Columns and mechanically agitated cells—A review. *Chem. Eng. Res. Des.* **2007**, *85*, 1591–1603. [CrossRef]
6. Cheng, G.; Cao, Y.; Zhang, C.; Jiang, Z.; Yu, Y.; Mohanty, M.K. Application of novel flotation systems to fine coal cleaning. *Int. J. Coal Prep. Util.* **2020**, *40*, 24–36. [CrossRef]
7. Moys, M.; Engelbrecht, J. Simulation of the behaviour of flexible baffles in flotation columns. *Chem. Eng. J. Biochem. Eng. J.* **1995**, *59*, 33–38. [CrossRef]
8. Vashisth, S.; Bennington, C.P.; Grace, J.R.; Kerekes, R.J. Column flotation deinking: State-of-the-art and opportunities. *Resour. Conserv. Recycl.* **2011**, *55*, 1154–1177. [CrossRef]
9. Ding, Y.; Wu, Y.; Li, D.; Zheng, J. Technical note a study on the mixing characteristics of a packed flotation column. *Miner. Eng.* **2001**, *14*, 1101–1105. [CrossRef]
10. Xia, Y.; Peng, F.; Wolfe, E. CFD simulation of alleviation of fluid back mixing by baffles in bubble column. *Miner. Eng.* **2006**, *19*, 925–937. [CrossRef]
11. Farzanegan, A.; Khorasanizadeh, N.; Sheikhzadeh, G.A.; Khorasanizadeh, H. Laboratory and CFD investigations of the two-phase flow behavior in flotation columns equipped with vertical baffle. *Int. J. Miner. Process.* **2017**, *166*, 79–88. [CrossRef]
12. Zhang, M.; Li, T.; Wang, G. A CFD study of the flow characteristics in a packed flotation column: Implications for flotation recovery improvement. *Int. J. Miner. Process.* **2017**, *159*, 60–68. [CrossRef]
13. Zhang, M.; Li, T.; Ma, S.; Wang, G. An experimental study of copper sulfide flotation in a packed cyclonic–static microbubble flotation column. *Sep. Sci. Technol.* **2018**, *53*, 2238–2248. [CrossRef]
14. Yan, X.; Shi, R.; Xu, Y.; Wang, A.; Liu, Y.; Wang, L.; Cao, Y. Bubble behaviors in a lab-scale cyclonic-static micro-bubble flotation column. *Asia-Pacific J. Chem. Eng.* **2016**, *11*, 939–948. [CrossRef]

15. Zhang, P.; Zhang, W.; Ou, L.; Zhu, Y.; Zhu, Z. Enhanced bauxite recovery using a flotation column packed with multilayers of medium. *Minerials* **2020**, *10*, 594. [CrossRef]

16. Luo, H.; Svendsen, H.F. Theoretical model for drop and bubble breakup in turbulent dispersions. *AIChE J.* **1996**, *42*, 1225–1233. [CrossRef]

17. Laari, A.; Turunen, I. Experimental determination of bubble coalescence and break-up rates in a bubble column reactor. *Can. J. Chem. Eng.* **2003**, *81*, 395–401. [CrossRef]

18. Sattar, M.; Naser, J.; Brooks, G. Numerical simulation of two-phase flow with bubble break-up and coalescence coupled with population balance modeling. *Chem. Eng. Process.* **2013**, *70*, 66–76. [CrossRef]

19. Lehr, F.; Millies, M.; Mewes, D. Bubble-size distributions and flow fields in bubble columns. *AIChE J.* **2002**, *48*, 2426–2443. [CrossRef]

20. Tao, D.; Luttrell, G.; Yoon, R.-H. A parametric study of froth stability and its effect on column flotation of fine particles. *Int. J. Miner. Process.* **2000**, *59*, 25–43. [CrossRef]

21. Yoon, R.H.; Luttrell, G.H. The effect of bubble size on fine particle flotation. *Miner. Process. Extr. Metall. Rev.* **1989**, *5*, 101–122. [CrossRef]

22. Schubert, H. On the turbulence-controlled microprocesses in flotation machines. *Int. J. Miner. Process.* **1999**, *56*, 257–276. [CrossRef]

23. Schubert, H.; Bischofberger, C. On the microprocesses air dispersion and particle-bubble attachment in flotation machines as well as consequences for the scale-up of macroprocesses. *Int. J. Miner. Process.* **1998**, *52*, 245–259. [CrossRef]

24. Koh, P.; Schwarz, M. CFD modelling of bubble-particle attachments in flotation cells. *Miner. Eng.* **2006**, *19*, 619–626. [CrossRef]

Article

Flotation Performance, Structure-Activity Relationship and Adsorption Mechanism of O-Isopropyl-N-Ethyl Thionocarbamate Collector for Elemental Sulfur in a High-Sulfur Residue

Guiqing Liu [1,2], Bangsheng Zhang [2], Zhonglin Dong [3,*], Fan Zhang [2], Fang Wang [2], Tao Jiang [3] and Bin Xu [3,*]

1   School of Metallurgy, Northeastern University, Shenyang 110819, China; Charles_liu32@163.com
2   Jiangsu BGRIMM Metal Recycling Science & Technology Co. Ltd., Xuzhou 221121, China; zbsvictory@163.com (B.Z.); bkyzhangfan@126.com (F.Z.); wangfang224444@126.com (F.W.)
3   School of Minerals Processing and Bioengineering, Central South University, Changsha 410000, China; jiangtao@csu.edu.cn
*   Correspondence: dongzhonglincsu@csu.edu.cn (Z.D.); xubincsu@csu.edu.cn (B.X.); Tel.: +86-183-7315-0200 (Z.D.); +86-150-8493-3770 (B.X.)

Citation: Liu, G.; Zhang, B.; Dong, Z.; Zhang, F.; Wang, F.; Jiang, T.; Xu, B. Flotation Performance, Structure-Activity Relationship and Adsorption Mechanism of O-Isopropyl-N-Ethyl Thionocarbamate Collector for Elemental Sulfur in a High-Sulfur Residue. *Metals* **2021**, *11*, 727. https://doi.org/10.3390/met 11050727

Academic Editors: Ilhwan Park and Anna H. Kaksonen

Received: 11 April 2021
Accepted: 25 April 2021
Published: 28 April 2021

Publisher's Note: MDPI stays neutral with regard to jurisdictional claims in published maps and institutional affiliations.

Abstract: O-isopropyl-N-ethyl thionocarbamate (IPETC) collector was used to selectively recover elemental sulfur from a high-sulfur residue, and its flotation performance, structure–property relationship and adsorption mechanism to elemental sulfur were studied. The raw ore flotation test showed that IPETC displayed superior flotation performance to the elemental sulfur compared with sodium ethyl xanthate (SEX) and ammonium dibutyl dithiophosphate (ADDTP) collectors. Pure mineral flotation and adsorption experiments further demonstrated that among the three collectors, IPETC had the strongest collecting power and the optimum selectivity towards elemental sulfur. The structure–property relationship research based on density functional theory (DFT) calculation supported the above conclusion. The adsorption mechanism analysis manifested that IPETC adsorption on elemental sulfur surface was a chemical process by separately generating normal covalent bond between carbonyl S atom and S atom and a backdonation covalent bond between O atom and S atom, which was confirmed by the FTIR spectrum analysis result. IPETC exhibits excellent collecting ability and selectivity for elemental sulfur and therefore it has bright application prospects.

Keywords: O-isopropyl-N-ethyl thionocarbamate (IPETC); elemental sulfur; flotation performance; structure–property relationship; adsorption mechanism

1. Introduction

Elemental sulfur is an important chemical material and has been widely used for producing sulfuric acid and other products such as fertilizers, matches, food preservation agents, carbon disulfide, cement, gun powder, surfactants, detergents, pharmaceuticals, pesticides and vulcanized rubbers [1–7]. Elemental sulfur is usually the by-product of hydrometallurgical leaching of nonferrous sulfide ores. For example, oxygen pressure acid leaching of zinc concentrate (sphalerite) can yield large amounts of elemental sulfur [8–14], and the overall reaction is shown in Equation (1).

$$ZnS + H_2SO_4 + 0.5O_2 = ZnSO_4 + H_2O + S \tag{1}$$

As indicated, zinc in sphalerite is transformed into zinc sulfate which can be used to produce zinc metal through electrowinning and sulfur remains in the residue in the form of elemental sulfur [15–18]. The residue is considered to be acid-producing since its sulfur component can react with oxygen and water to generate thiosalts and sulfuric acid [19]. At present, the produced high-sulfur residue is usually subjected to safe storage for most

Metals **2021**, *11*, 727. https://doi.org/10.3390/met11050727   https://www.mdpi.com/journal/metals

zinc refineries in China, which not only causes serious resource waste but also poses huge threat to the environment and human health [20,21]. Therefore, it is necessary to recover elemental sulfur from the residue with effective technology.

Elemental sulfur has good natural hydrophobicity, and therefore without the use of any additional agent, it can be efficiently enriched by froth flotation, an extensively used technology in mineral processing for separating target mineral from gangue mineral by using their natural hydrophobicity difference [22–25]. However, the nonferrous metal sulfides (ZnS, FeS_2, PbS and Ag_2S) that are not completely oxidized during pressure leaching also remain in the residue [26], and inevitably they will float upward together with elemental sulfur and enter the concentrate because of their similar floatability and intimate association relationship [27,28]. This leads to the decrease of elemental sulfur grade in the concentrate product. To enhance the grade, a hot filtration procedure is usually implemented for the concentrate, which can realize the efficient separation of elemental sulfur from the concentrate by melting and filtration at 115–155 °C [19]. Nevertheless, a significant portion of the elemental sulfur is still hard to separate from the concentrate with hot filtration since this process generally needs raw material whose elemental sulfur content is more than 70% to ensure a satisfactory performance [26]. The elemental sulfur content in most of the high-sulfur residue is usually in 40–60% [15], and thus it is difficult to obtain a high elemental sulfur recovery. A feasible method to solve the above problem is to add suitable collector during the flotation procedure to realize the selective flotation of elemental sulfur.

The study aims to find an efficient collector to selectively separate elemental sulfur from a high-sulfur pressure acid leaching residue of zinc sulfide concentrate. First, the flotation performance of O-Isopropyl-N-Ethyl thionocarbamate (IPETC), ammonium dibutyl dithiophosphate (ADDTP) and sodium ethyl xanthate (SEX) for elemental sulfur were compared through a raw ore flotation experiment. Then, pure mineral flotation and adsorption tests were performed to further compare their flotation properties. After that, the structure-activity relationships of three collectors were researched using density functional theory (DFT) calculation through comparing their geometric configurations, electronic structures and corresponding flotation performances. Based on the above results, IPETC was selected as the optimum elemental sulfur collector, and its adsorption mechanism on the elemental sulfur surface was investigated using adsorption simulation calculation and Fourier-transform infrared spectroscopy (FTIR) analysis.

2. Experimental

2.1. Material and Reagents

The pressure acid leaching residue of zinc sulfide concentrate used in this study was provided by Hulun Buir Chihong Mining Industry Co., Inner Mongolia, China, and it has been used in our previous work [15]. The chemical composition analysis showed that the sulfur content arrived at 46.21%, and 81.97% of the sulfur occurred in the form of elemental sulfur. Therefore, it was a high-sulfur residue and used for closed-circuit flotation experiment to recover elemental sulfur.

Previous literature showed that natural minerals and minerals after leaching have similar surface properties [29,30], and thus natural minerals are usually adopted as the raw materials for mineral flotation study. According to the result of the X-ray diffraction (XRD) spectra analysis [15], elemental sulfur, sphalerite, pyrite, chalcopyrite, albite and anglesite were the main minerals in the residue, and therefore these pure mineral samples obtained from Guangdong Province in China were chosen for the pure mineral flotation experiment. The high-purity minerals were first collected by handpicking, and then crushed and ground in a porcelain ball mill. Afterwards, the samples were dry-sieved to obtain the $-74 + 38$ μm fractions for the flotation experiment, and the fractions of less than 5 μm were used for FTIR spectrum measurement. The X-ray fluorescence (XRF) analysis results for the samples in Tables 1–3 showed that their purities were over 96%, which could meet the desirable requirement in this study.

Table 1. Chemical compositions of pure minerals of elemental sulfur, sphalerite, pyrite and chalcopyrite (wt %).

Mineral	Cu	Pb	Zn	Fe	S	Purity
Elemental sulfur	——	——	—	0.02	98.68	98.68
Sphalerite	—	0.04	66.20	0.52	32.9	98.65
Pyrite	—	0.21	0.07	45.25	53.26	97.21
Chalcopyrite	34.13	0.01	0.27	30.10	34.22	98.76

Note: "——" indicates that the element was not detected.

Table 2. Chemical composition of pure albite mineral (wt %).

Mineral	Na_2O	Al_2O_3	SiO_2	Purity
Albite	11.36	19.19	67.86	96.27

Table 3. Chemical composition of pure anglesite mineral (wt %).

Mineral	PbO	SO_3	Fe_2O_3	Purity
Anglesite	69.32	28.23	0.052	96.27

In order to investigate the difference of flotation property of pure minerals and pure minerals after oxygen pressure acid leaching, a comparative experiment was conducted, and the result is shown in Table 4. The method of the flotation test was the same as that in Section Pure Mineral Flotation, and the collector (IPETC) dosage and pulp pH value are 3×10^{-5} mol/L and 8.0, respectively. The recoveries of pure minerals of elemental sulfur, sphalerite, pyrite, chalcopyrite, albite and anglesite were 99.2%, 51.6%, 78.3%, 82.5%, 23.6% and 13.3%, respectively, while the recoveries of the above pure minerals after leaching were 99.0%, 51.3%, 78.2%, 82.1%, 23.4% and 13.5%, respectively. Thus, a very close flotation recovery was obtained for each pure mineral and pure mineral after leaching, which can be ascribed to their close floatability. Therefore, it can be deduced that the physical and chemical properties of the surfaces of the pure minerals and pure minerals after leaching are very close.

Table 4. The flotation results of pure minerals and pure minerals after oxygen pressure acid leaching whose conditions are the following: initial sulfuric acid concentration, 60 g/L, temperature, 150 °C, oxygen partial pressure, 1.0 MPa and time, 2.5 h.

Raw Material	Recovery (%)					
	Elemental Sulfur	Sphalerite	Pyrite	Chalcopyrite	Anglesite	Albite
Pure minerals	99.2	51.6	78.3	82.5	23.6	13.3
Pure minerals after oxygen pressure acid leaching	99.0	51.3	78.2	82.1	23.4	13.5

The analytically pure O-Isopropyl-N-Ethyl thionocarbamate (IPETC), ammonium dibutyl dithiophosphate (ADDTP) and sodium ethyl xanthate (SEX) were bought from a commercial company of Hunan Province, China, and used as the collector. Methyl isobutyl carbinol (MIBC) from the same company was used as the foaming agent. The other reagents used are also analytically pure. Ultrapure water was used throughout all experiments.

2.2. Experimental Methods

2.2.1. Flotation Experiment

Raw Ore Flotation

The flotation experiment of pressure acid leaching residue was carried out in a self-aeration XFD-63 flotation machine from Jilin Prospecting Machinery Factory, China [31]. The residue and ultrapure water were first put into the cell to form the pulp with 25%

density, and then slime was added to adjust the pulp pH to about 8.0. Afterwards, the inhibitor (Na_2S + Na_2SO_3 + $ZnSO_4$), collector and foaming agent (MIBC) were successively added into the pulp which was agitated at 1650 rpm for 3 min after adding each reagent. The flotation was conducted for 5 min, and elemental sulfur concentrate and tailing were collected, filtrated, dried, weighed and assayed to calculate elemental sulfur recovery.

Pure Mineral Flotation

The pure mineral flotation experiment was performed in an XFG5-35 flotation machine (Jilin Prospecting Machinery Factory, Jilin, china) with a 40 mL plexiglass cell [32]. The agitating speed was kept at 1650 r min^{-1} with a mechanical impeller. In each test, 2.0 g pure mineral sample was placed into an ultrasonic bath for 5 min of cleaning, and then transferred into the flotation cell. After that, the pulp pH was adjusted to a desired value with dilute NaOH or H_2SO_4 solution. Finally, the collector and the foaming agent were sequentially added into the pulp which was agitated for 2 min. The flotation was carried out for 5 min, and the obtained concentrate and tailing were weighted and analyzed after filtration and drying to calculate mineral recovery.

2.2.2. FTIR Spectrum

IPETC solution (1×10^{-2} mol/L) was first mixed with elemental sulfur sample (0.5 g, -38 μm). After magnetically stirring for 60 min, the solution was filtered, and the obtained solid product was dried at 35 °C under vacuum for 24 h for subsequent infrared detection. The Fourier-transform infrared (FTIR) spectra were recorded by a G510PFTIR infrared spectrometer (Nicolet Company, Madison, USA) in the wavenumber range from 400 cm^{-1} to 4000 cm^{-1} using the KBr technique.

2.2.3. DFT Calculation

The initial molecular/ion model of collector was constructed and then optimized by MM2 and PM3 methods in Chemoffice 2008 program. The obtained geometry was further optimized by DFT method at the B3LYP/6-31G (d, p) level in the Gaussian 03 program [31]. Finally, periodic structure optimization was completed in the CASTEP module of Materials Studio 4.4 and then the optimized structure was subjected to DFT calculations.

The calculations of mineral plane and crystal were completed by CASTEP and Dmol3 modules in Materials Studio 4.4 program, in which CASTEP module was used for the establishment of plane and crystal models as well as geometric structure optimization, state density analysis and Mulliken population analysis, and Dmol3 module was used to analyze frontier orbital energy. The plane wave basis set and DND basis set were separately implemented in the CASTEP and Dmol3 modules for the DFT calculations. The establishment and optimization of adsorption model of IPETC on elemental sulfur surface and related calculation were both performed in the CASTEP module of Materials Studio 4.4.

2.3. Analytical Methods

The elemental sulfur content in the high-sulfur residue and flotation concentrate was detected by carbon tetrachloride dissolution followed by gravimetric analysis [33]. The concentrations of sulfhydryl collectors were determined by UV-Vis spectroscopy [34].

3. Results and Discussion

3.1. Flotation Performance of Collectors

Before the use of three collectors IPETC, ADDTP and SEX, single foaming agent MIBC and non-polar collector kerosene were separately used for recovering elemental sulfur from the high-sulfur residue. The flotation sheets are displayed in Figure 1a,b, and the flotation results are shown in Table 5. It can be seen that a low and close elemental sulfur recovery was obtained for MIBC and kerosene. Furthermore, the elemental sulfur grades in the concentrate were low, which is because large amounts of sulfide minerals such as sphalerite, pyrite, chalcopyrite in the residue entered the concentrate with elemental sulfur.

The above results indicated that the selective and effective separation of elemental sulfur from the high-sulfur residue cannot be realized using MIBC and kerosene.

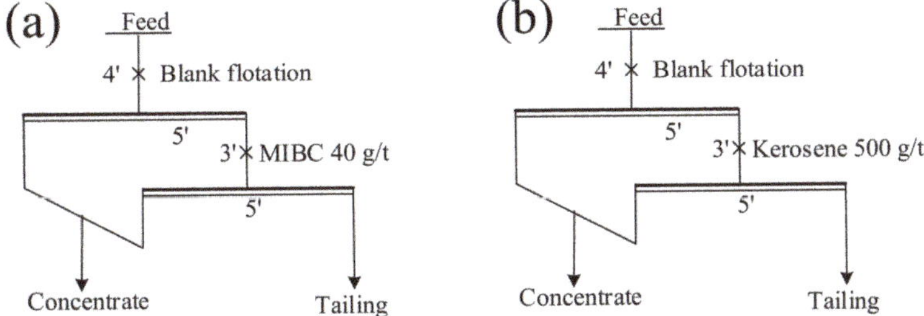

Figure 1. The flotation sheets of high-sulfur residue with MIBC (**a**) and kerosene (**b**).

Table 5. The flotation results of high-sulfur residue with MIBC and kerosene.

Collector	Product	Yield (%)	Elemental Sulfur Grade (%)	Elemental Sulfur Recovery (%)
MIBC	Elemental sulfur concentrate	57.31	71.17	88.83
	Tailing	42.69	12.02	11.17
	Feed residue	100.00	45.92	100.00
Kerosene	Elemental sulfur concentrate	56.70	71.24	88.00
	Tailing	43.30	12.72	12.00
	Feed residue	100.00	45.90	100.00

The flotation performances of IPETC, ADDTP and SEX collectors for elemental sulfur in the high-sulfur residue were compared. After detailed conditional experiment in our laboratory, the closed-circuit flotation flowchart in Figure 2 was ascertained and adopted, and the obtained results are listed in Table 6. Despite the high pulp pH, the collectors will also adsorb a small amount of sphalerite. Thus, zinc sulfate is added to increase the inhibitory effect on sphalerite, even though some zinc ions are present in the pulp itself. It is not necessary to inhibit the gangue minerals in the residue, such as albite and anglesite, because they have poor natural floatability and the sulfhydryl collector (i.e., IPETC) hardly adsorbs on their surfaces.

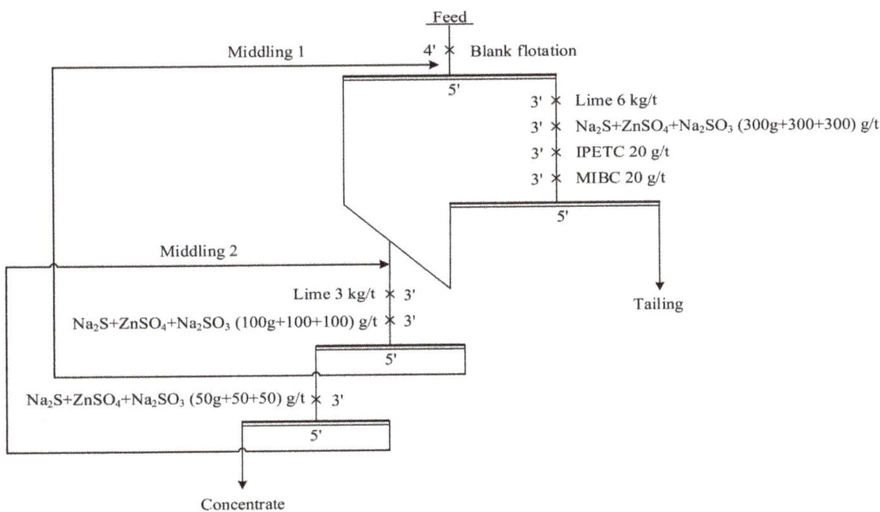

Figure 2. Flowchart of closed-circuit flotation of high-sulfur residue with IPETC, ADDTP and SEX collectors.

Table 6. The result of closed-circuit flotation of the high-sulfur residue with IPETC, ADDTP and SEX collectors.

Collector	Product	Yield (%)	Grade (%)		Recovery (%)	
			Elemental Sulfur	**Zinc**	**Elemental Sulfur**	**Zinc**
SEX	Elemental sulfur concentrate	43.55	81.53	1.33	93.67	13.33
	Tailing	56.45	4.25	6.67	6.33	86.67
	Feed residue	100	37.91	4.34	100	100
ADDTP	Elemental sulfur concentrate	45.47	82.25	1.15	97.83	12.08
	Tailing	54.53	1.52	6.98	2.17	87.92
	Feed residue	100	38.23	4.33	100	100
IPETC	Elemental sulfur concentrate	44.78	84.47	0.88	99.87	9.06
	Tailing	55.22	0.09	7.16	0.13	90.94
	Feed residue	100	37.88	4.35	100	100

From Table 6, the elemental sulfur recoveries for SEX, ADDTP and IPETC were 93.67%, 97.83% and 99.87% while their elemental sulfur grades in the concentrate were 81.53%, 82.25% and 84.47%, respectively. Thus, both the maximum elemental sulfur recovery and grade were achieved using IPETC collector. The zinc flotation result was also shown in Table 6. The zinc grades in the elemental sulfur concentrate were 1.33%, 1.15% and 0.88% for SEX, ADDTP and IPETC, and their recoveries were 13.33%, 12.08% and 9.06%, respectively. Thus, compared with elemental sulfur, only a small amount of zinc was floated and entered the concentrate. Moreover, the zinc grade in the concentrate and its recovery were the lowest for IPETC. The above results indicated that in comparison with SEX and ADDTP, IPETC exhibited more excellent collecting ability and selectivity for elemental sulfur in the high-sulfur residue, and thus it has good industrial application potential.

3.2. Pure Mineral Flotation and Adsorption Experiments

3.2.1. Pure Mineral Flotation Experiment

The collecting performances of IPETC, ADDTP and SEX for pure minerals of elemental sulfur, sphalerite, pyrite, chalcopyrite, albite and anglesite were compared, and the results of effects of collector dosage and pulp pH on pure mineral recovery are displayed in Figures 3a–c and 4a–c.

As shown in Figure 3a–c, the recoveries of elemental sulfur, sphalerite, pyrite and chalcopyrite went up with the increase of IPETC, ADDTP and SEX dosages in their initial ranges of $(2$–$3) \times 10^{-5}$, $(0.4$–$1.2) \times 10^{-4}$ and $(0.4$–$1.2) \times 10^{-4}$ mol/L. Afterwards, no obvious increases of the recoveries of four pure minerals were observed with the further increase of collector dosage. Therefore, the optimal IPETC, ADDTP and SEX dosages were separately 3×10^{-5}, 1.2×10^{-4} and 1.2×10^{-4} mol/L. The recoveries of albite and anglesite were low in the whole abscissa range, and collector dosage had little effect on them, which could be ascribed to their poor natural floatability.

As indicated in Figure 4a–c, the recoveries of elemental sulfur and three sulfides were all augmented as the pulp pH increased in the initial range of 2–8. Nevertheless, further increase of pulp pH from 8 to 12 resulted in the decrease of the recoveries, but the decrease degrees of recoveries of three sulfides were larger than that of elemental sulfur, suggesting that increased pH had less negative impact on elemental sulfur flotation in this pH range. Thus, the optimal pulp pH for the four minerals was 8. The pulp pH exerted similar effect on anglesite recovery, but its optimum pH was 6 where the maximum recovery was achieved. The albite recovery gradually dropped in the pulp pH range of 2–8 and then basically remained steady.

In addition, for each collector, both the maximum elemental sulfur recovery and the differences between the maximum recoveries of elemental sulfur and the other five pure minerals followed the order of IPETC > ADDTP > SDD. Therefore, among the three collectors, IPETC presented the optimal collecting ability and selectivity to elemental sulfur.

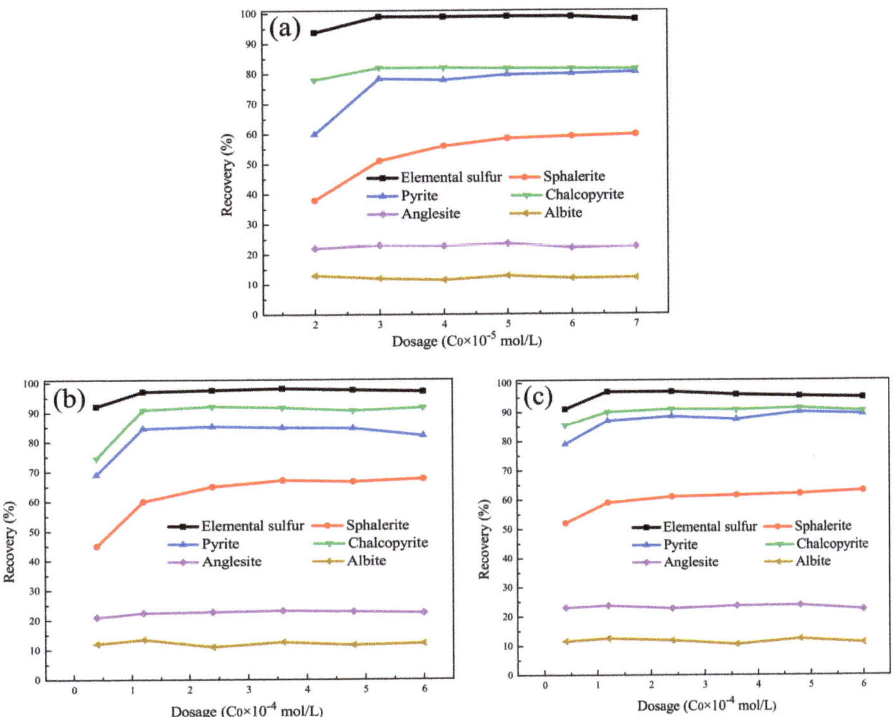

Figure 3. Effect of collector dosage on pure mineral recovery at pulp pH value 8 ((**a**) IPETC; (**b**) ADDTP; (**c**) SEX).

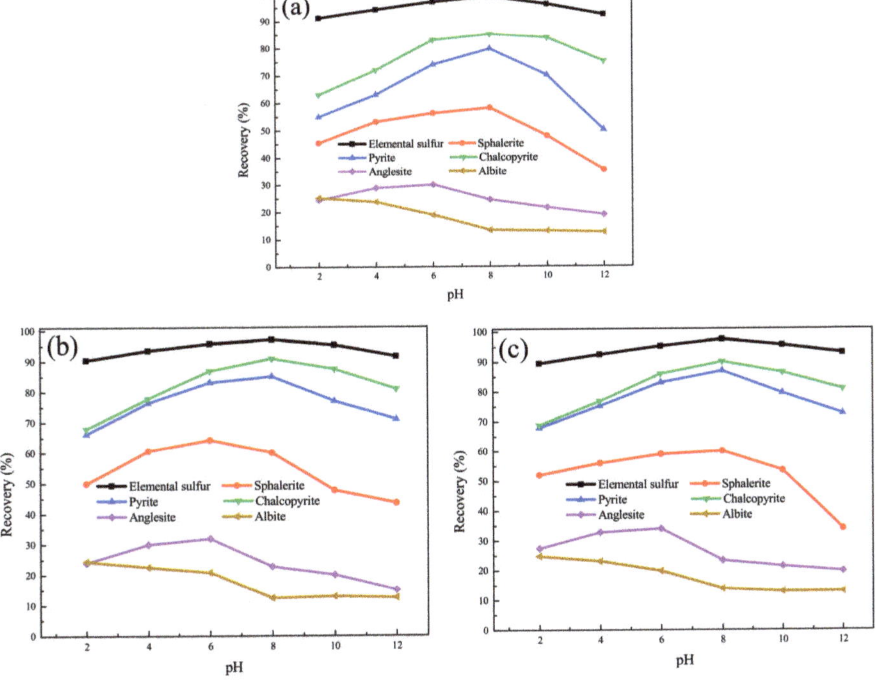

Figure 4. Effect of pulp pH on pure mineral recovery at optimum collector dosage ((**a**) IPETC; (**b**) ADDTP; (**c**) SEX).

3.2.2. Pure Mineral Adsorption Experiment

The adsorption amounts of IPETC, ADDTP and SEX on pure mineral surfaces were also compared, and the effect of initial collector concentration is shown in Figure 5a–c. With the increase of concentrations of three collectors, their adsorption amounts on the surfaces of elemental sulfur, albite and anglesite basically remained steady, but the adsorption amount on elemental sulfur surface kept at a high level while those on the surfaces of albite and anglesite were very low. For sphalerite, pyrite and chalcopyrite, the adsorption amounts of three collectors on their surfaces first increased and then were roughly unchanged.

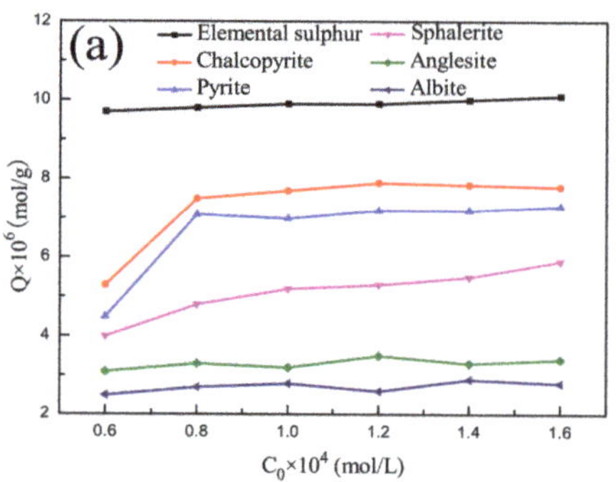

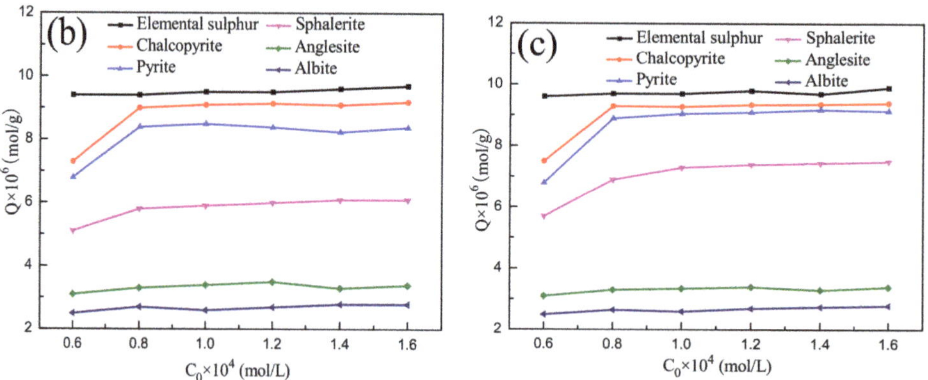

Figure 5. Effect of initial collector concentration on their adsorption amounts on pure mineral surfaces at pulp pH 8 ((**a**) IPETC; (**b**) ADDTP; (**c**) SEX).

However, among the three collectors, the adsorption amount of IPETC on elemental sulfur surface was the biggest under the same collector concentration, suggesting that IPETC possessed the strongest interaction with elemental sulfur surface. Moreover, IPETC exhibited the greatest difference between adsorption quantities on elemental sulfur surface and the surfaces of other pure minerals. The above results indicated that IPETC not only possessed the biggest adsorption amount on elemental sulfur surface but also displayed the best adsorption selectivity for elemental sulfur, which is consistent with the result of the pure mineral flotation test in Section 3.2.1.

3.3. Structure–Property Relationships of Collectors

3.3.1. Geometry Configuration

The simulation calculations of collector molecules/ion were carried to study their structure–property relationships. The optimized geometry configurations of ADDTP, SEX and IPETC are presented in Figure 6a–c, and the bond lengths and dihedral angles are displayed in Table 7.

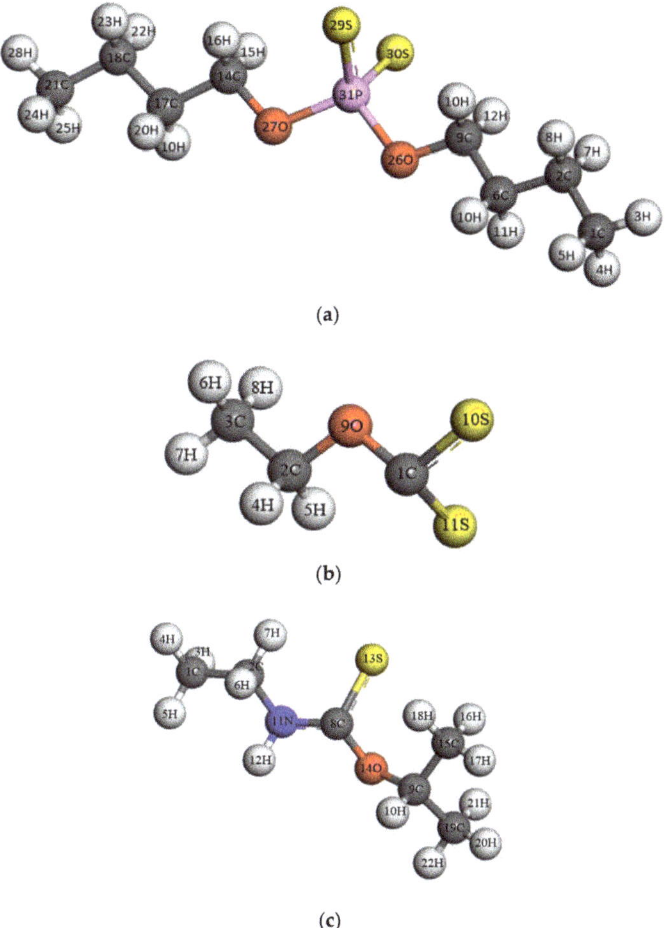

(a)

(b)

(c)

Figure 6. Optimized geometry configurations of collector molecules or ions ((**a**) IPETC; (**b**) ADDTP; (**c**) SEX).

Table 7. Bond lengths and dihedral angles of collectors.

Collector	Bond Length (Å)		Dihedral Angle (°)	
SEX	1C-10S	1C-11S	9O-1C-10S-11S	
	1.707	1.699	−179.992	
ADDTP	29S-31P	30S-31P	14C-27O-31P-29S	9C-26O-31P-30S
	2.006	2.006	−59.008	−59.003
IPETC	8C-13S		13S-8C-11N-2C	11N-8C-14O-9C
	1.679		−178.654	177.847

The bond length of carbonyl S atom in SEX is longer than that in IPETC, indicating that carbonyl S in IPETC is less likely to lose electrons. As a result of this, normal covalent bond is easier to be generated in SEX, which is not unbeneficial to its selectivity. Therefore,

IPETC presents a better collecting selectivity for elemental sulfur in terms of bond length of carbonyl S atom. The carbonyl S atom in ADDTP bonds with the P atom to form a P=S bond which is different with C=S bond in SEX and IPETC, and thus their bond lengths are not comparable.

Judging from the dihedral angle of three collectors, the atoms of -S-C=N-C- and -N-C=O-C- groups in IPETC and -O-C(=S)-S- group in SEX are nearly located in the same plane, respectively. Furthermore, the p orbitals that are not involved in hybridization exist in the valence electron layers of these atoms. Thus, conjugated big π-bond is easy to be formed in the three groups.

3.3.2. Electronic Structure

Mulliken Population Analysis

The Mulliken populations of bond and atom charge of SEX, ADDTP and IPETC are indicated in Table 8. The sequence of bond populations of C=S or P=S bonds in the three collectors is IPETC > ADDTP > SEX. A larger population means stronger covalency of the covalent bond. Thus, in comparison with IPETC, it is more likely for the carbonyl S atoms in ADDTP and SEX to offer electrons, leading to the easier generation of a normal covalent bond between collector and mineral surface. Therefore, in terms of the bond population, IPETC shows better selectivity to elemental sulfur.

Table 8. Mulliken populations of C=S or P=S bond and atom charge of collectors.

Collector	Bond Population		Atom Charge Population			
SEX	10S-1C 0.41	11S-1C 0.31	1C −0.099	9O −0.174	10S −0.379	11S −0.405
ADDTP	29S-31P 0.72	30S-31P 0.74	27O −0.729	29S −0.759	30S −0.811	31P 1.521
IPETC	8C-13S 1.00		8C 0.067	11N 0.022	13S −0.206	14O −0.238

The absolute values of Mulliken charge populations of carbonyl S atoms follow the order of ADDTP > SEX > IPETC. A larger absolute value means stronger electrostatic attraction between carbonyl S atom in collector and S atom in elemental sulfur surface. Nevertheless, electrostatic interaction has no directivity, and therefore the larger absolute value is unfavorable to the collector selectivity. Thus, in terms of atom charge population, IPETC also exhibits better selectivity for elemental sulfur than ADDTP and SEX.

Density of States Analysis

The densities of states of three collectors are indicated in Figure 7a–c. The valence band tops of SEX, ADDTP and IPETC are mainly comprised of 3p orbitals of 10S and 11S, 29S and 30S, and 13S atoms, respectively. Their conduction band bottoms are primarily comprised of 2p orbitals of 1C and 9O atoms and 3p orbitals of 10S and 11S atoms, 2p orbitals of 26O, 27O and 31P atoms and 3p orbitals of 29S and 30S atoms, and 2p orbitals of 8C, 11N and 14O atoms and 3p orbital of 13S atom, respectively. According to the band theory, the valence band top is the highest occupied molecular orbital (HOMO), and the conduction band bottom is the lowest unoccupied molecular orbital (LUMO), and they have the highest chemical activity and are separately apt to lose and gain electrons. Therefore, it can be concluded that the oxidation reaction centers of SEX, ADDTP and IPETC are separately 10S and 11S, 29S and 30S, and 13S atoms. Accordingly, the reduction reactions can occur on 1C, 9O, 10S and 11S atoms for SEX, 26O, 27O, 29S, 30S and 31P atoms for ADDTP, and 8C, 11N, 13S and 14O atoms for IPETC, respectively.

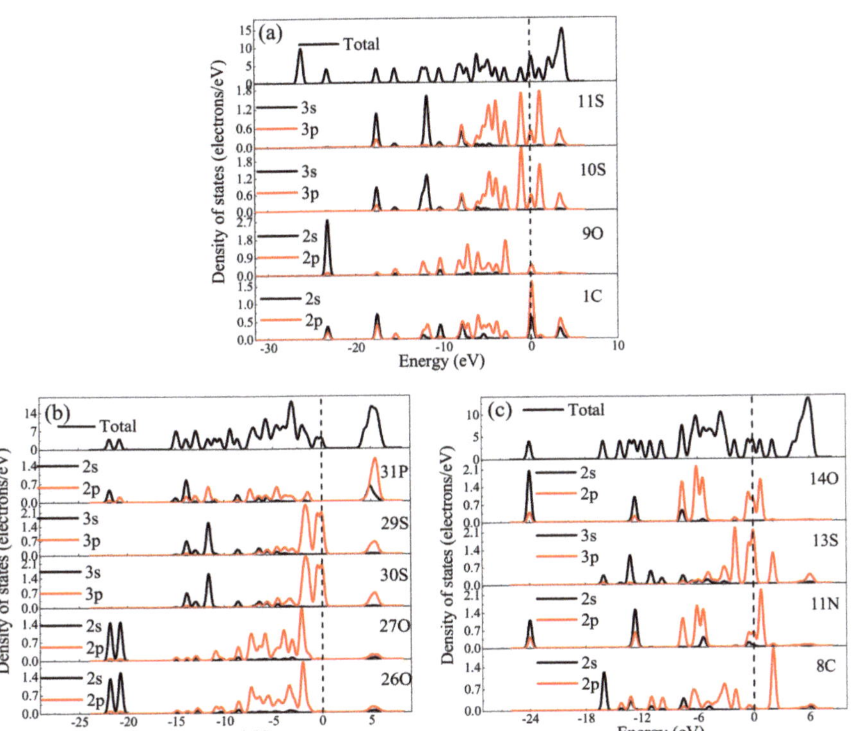

Figure 7. Densities of states of collectors ((**a**) IPETC; (**b**) ADDTP; (**c**) SEX).

Frontier Orbital Analysis

The electronic cloud pictures of frontier molecular orbitals of three optimized SEX, ADDTP and IPETC models are given in Figure 8a–f. From these subgraphs, the HOMOs of SEX, ADDTP and IPETC are mainly composed of electron orbitals of 10S and 11S, 29S and 30S, and 13S atoms, respectively. The LUMOs are primarily constituted by electron orbitals of 1C, 9O, 10S and 11S atoms for SEX, 26O, 27O, 29S, 30S and 31P atoms for ADDTP, and 8C, 11N, 13S and 14O atoms for IPETC, respectively. The above results are consistent with the results of density of states analysis in Section Density of States Analysis. Furthermore, the atoms of -S-C=N-C- and -N-C=O-C- groups in IPETC and -O-C(=S)-S- group in SEX are all almost in the same plane based on the dihedral angle values in Table 7. Moreover, it can be known from the result of density of states analysis in Section Density of States Analysis that the LUMO orbitals of IPETC and SEX all constitute the p orbital of each atom in these functional groups. Therefore, it can be inferred that the LUMO orbitals of IPETC and SEX both are conjugated big π-bonds formed by electron orbitals of these functional groups.

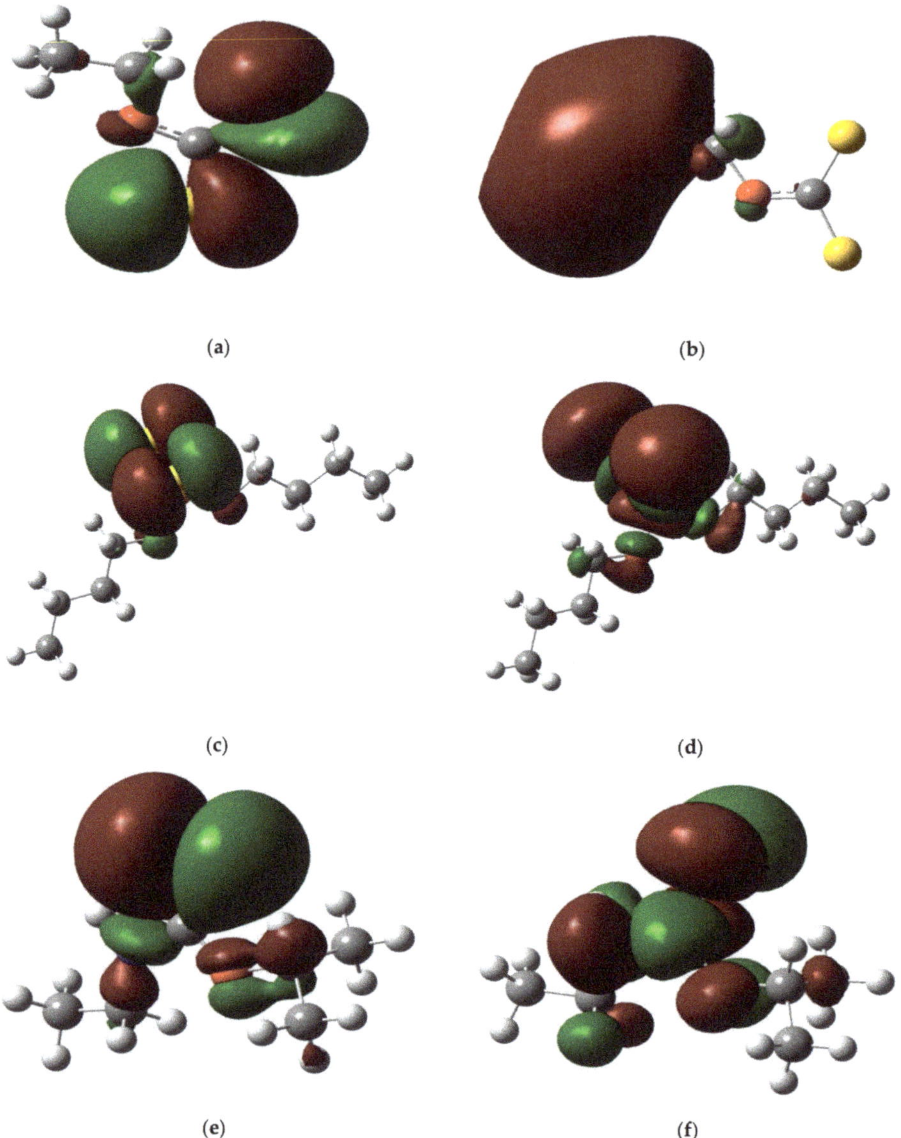

Figure 8. Electron cloud pictures of frontier molecular orbitals of optimized collector models ((**a**) HOMO of SEX; (**b**) LUMO of SEX; (**c**) HOMO of ADDTP; (**d**) LUMO of ADDTP; (**e**) HOMO of IPETC; (**f**) LUMO of IPETC).

The frontier orbital energies of three collectors and energy differences between collectors and minerals (ΔE) are shown in Table 9. According to frontier molecular orbital theory, a smaller absolute value of energy difference between the HOMO of one reactant and the LUMO of another one means that they are prone to react with each other. ΔE_1 are all evidently smaller than ΔE_2, suggesting that the chemical reaction between collectors and minerals is led by the electron transfer from collectors to minerals. Thus, the normal covalent bond predominates in the interaction. ΔE_1 between each collector and elemental sulfur is smaller than that between each collector and chalcopyrite/sphalerite/pyrite/anglesite/albite, indicating that the normal covalent bond between each collector and elemental sulfur is stronger than that between each collector and five other minerals. In comparison with SEX and ADDTP, both ΔE_1 between IPETC and elemental sulfur and ΔE_1 elemental sulfur $/\Delta E_1$ five other minerals of IPETC are the minimum. Thus, IPETC displays the optimal collecting power and selectivity for elemental sulfur among the three collectors.

Table 9. Frontier orbital energies of collectors and energy differences between collectors and minerals.

Collector	Orbital Energy (eV)		ΔE^a	Elemental Sulfur	Chalcopyrite	Pyrite	Sphalerite	Anglesite	Albite
SEX	HOMO	−4.57	ΔE_1	0.39	0.64	0.78	1.54	1.69	1.77
	LUMO	−3.07	ΔE_2	2.55	2.60	2.21	2.38	2.44	2.67
ADDTP	HOMO	−5.56	ΔE_1	0.32	0.51	0.53	0.73	0.95	1.23
	LUMO	−4.67	ΔE_2	2.42	1.00	1.37	2.04	2.31	2.45
IPETC	HOMO	−5.18	ΔE_1	0.15	0.76	1.04	1.43	1.51	1.76
	LUMO	−3.76	ΔE_2	1.91	1.86	1.77	1.91	1.97	2.12

$^a\Delta E_1 = |E(\text{HOMO, collector}) - E(\text{LUMO, mineral})|$ and $\Delta E_2 = |E(\text{HOMO, mineral}) - E(\text{LUMO, collector})|$.

3.4. Mechanism of IPETC Adsorption on Elemental Sulfur Surface

3.4.1. Adsorption Configuration and Adsorption Energy

After many adsorption position tests, six possible configurations of IPETC adsorption on perfect elemental sulfur (110) plane are indicated in Figure 9a–f, and the adsorption energies are shown in Table 10. The adsorption energies of six configurations are all negative, and therefore these adsorption reactions can happen spontaneously. Nevertheless, the adsorption energy is the lowest for the configuration of simultaneous adsorption of carbonyl S together with O, suggesting that this configuration is the most stable.

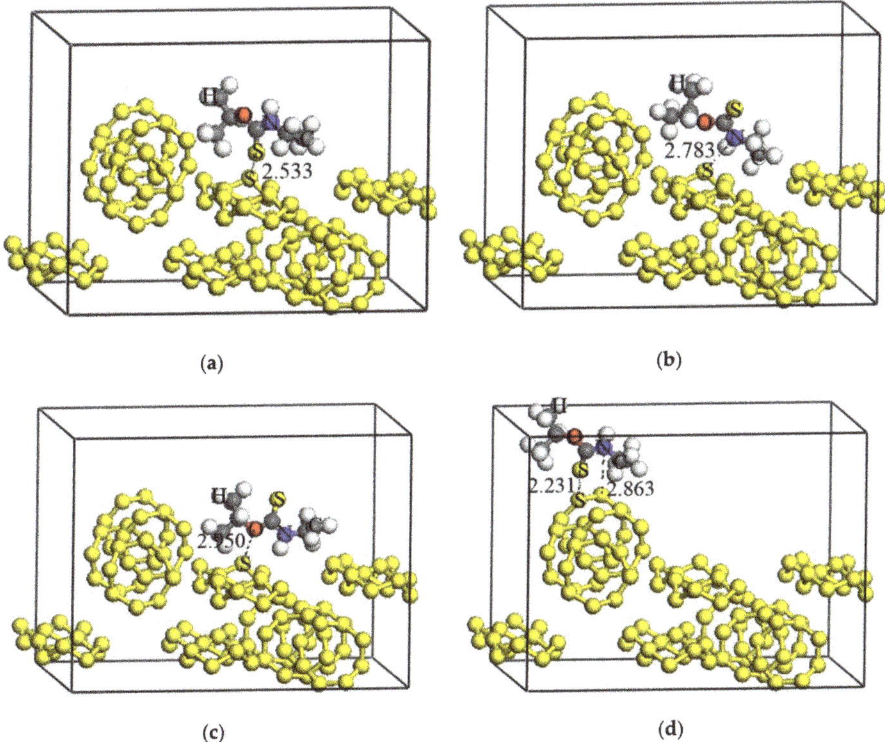

Figure 9. *Cont.*

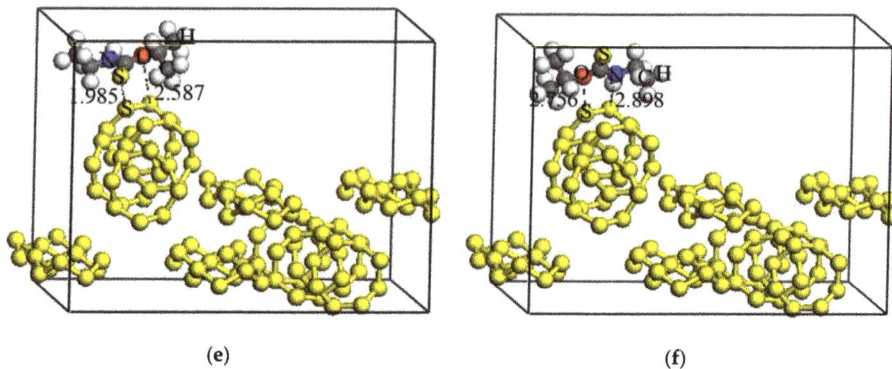

(e)   (f)

Figure 9. Adsorption configurations of IPETC on perfect elemental sulfur (100) crystal plane ((**a**) adsorption of carbonyl S atom; (**b**) adsorption of N atom; (**c**) adsorption of O atom; (**d**) adsorption of carbonyl S together with N atoms; (**e**) adsorption of carbonyl S together with O atoms; (**f**) adsorption of N together with O atoms).

Table 10. Adsorption energy of each adsorption configuration of IPETC on elemental sulfur (110) crystal plane.

Adsorption Configuration	Adsorption Energy (kJ/mol)
Adsorption of carbonyl S	−19.79
Adsorption of N	−10.23
Adsorption of O	−6.16
Simultaneous adsorption of carbonyl S together with N	−21.82
Simultaneous adsorption of carbonyl S together with O	−26.98
Simultaneous adsorption of N and O	−16.53

The adsorption energies of six configurations of IPETC adsorption on the perfect crystal planes of chalcopyrite (112), sphalerite (110), pyrite (100), anglesite (001) and albite (001) are displayed in Table 11. Obviously, the most stable adsorption configuration of IPETC on the five crystal planes is also the simultaneous adsorption of carbonyl S together with O. Nevertheless, the adsorption energies for the first four minerals are negative and obey the sequence of chalcopyrite < pyrite < sphalerite < anglesite, which are all larger than that for elemental sulfur in Table 9. Thus, it is easier for IPETC adsorption on elemental sulfur surface to occur. The adsorption energies for albite are all positive, and therefore IPETC adsorption on albite surface is difficult, which supported the results of pure mineral flotation and adsorption experiments in Section 3.2 that both the albite recovery and its adsorption amount on albite surface are very low.

Table 11. Adsorption energy of each adsorption configuration of IPETC on each crystal plane.

Adsorption Configuration	Adsorption Energy (kJ/mol)				
	Chalcopyrite (112) Plane	Pyrite (100) Plane	Sphalerite (110) Plane	Anglesite (001) Plane	Albite (001) Plane
Adsorption of carbonyl S	−12.53	−11.12	−10.02	−3.53	1.54
Adsorption of N	−6.88	−3.78	−2.12	−1.38	4.10
Adsorption of O	−9.12	−8.00	−7.79	−3.19	2.16
Simultaneous adsorption of carbonyl S together with N	−8.01	−4.56	−3.45	−2.41	2.33
Simultaneous adsorption of carbonyl S together with O	−16.89	−14.69	−12.21	−4.67	1.23
Simultaneous adsorption of N and O	−5.13	−6.69	−5.63	−0.43	3.49

The steadiest adsorption configurations of IPETC on perfect planes of chalcopyrite, pyrite, sphalerite and anglesite are presented in Figure 10a–d, and the Mulliken populations of corresponding adsorption bonds are shown in Table 12. For each configuration, the population of carbonyl S adsorption bond is much bigger than that of O adsorption bond. Thus, the interaction between carbonyl S and bonding atom in mineral surface is stronger.

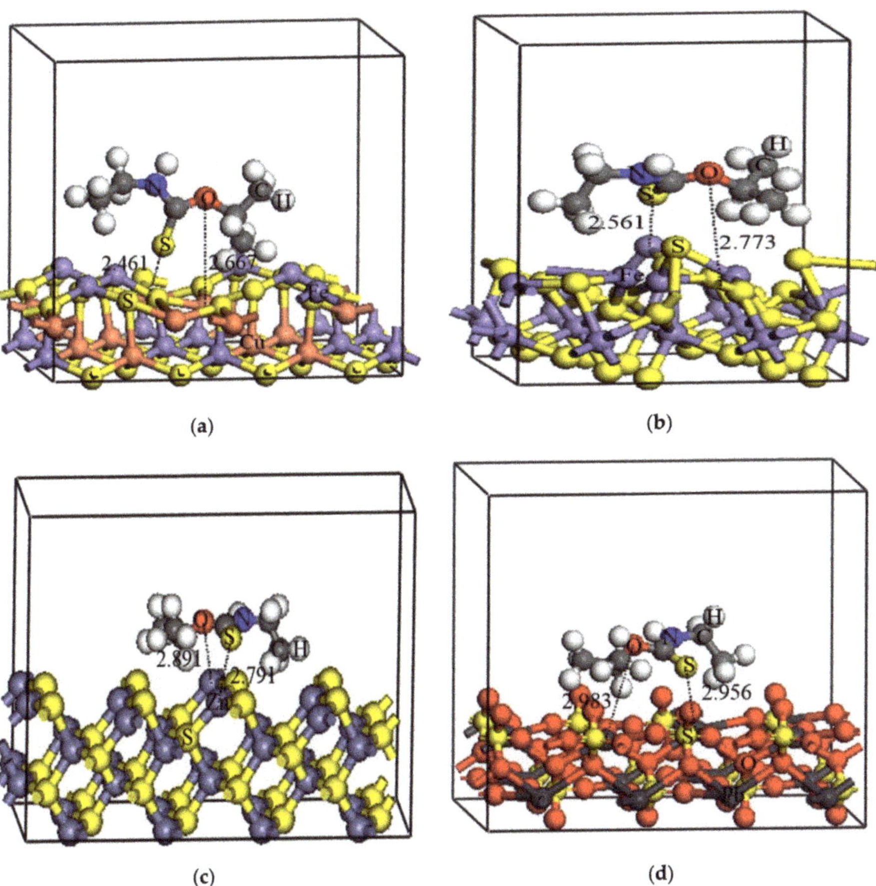

Figure 10. The steadiest adsorption configurations of IPETC on perfect crystal planes of chalcopyrite (112) (**a**), pyrite (100) (**b**), sphalerite (110) (**c**) and anglesite (001) (**d**).

Table 12. Mulliken population of adsorption bonds of each steadiest adsorption configuration of IPETC on each crystal plane.

Crystal Plane	Adsorption Bond	Bond Length (Å)	Mulliken Population
Elemental sulfur (110)	S-S	1.985	0.42
	O-S	2.587	0.15
Chalcopyrite (112)	S-Cu	2.461	0.35
	O-Cu	2.667	0.12
Pyrite (100)	S-Fe	2.561	0.33
	O-Fe	2.773	0.08
Sphalerite (110)	S-Zn	2.791	0.23
	O-Zn	2.891	0.11
Anglesite (001)	S-Pb	2.956	0.19
	O-Pb	2.983	0.12

3.4.2. Mulliken Population and Density of States Analyses

The Mulliken populations and densities of states of bonding atoms before and after IPETC adsorption on perfect elemental sulfur (100) plane are presented in Table 13 and Figure 11, respectively. The 3p orbital of carbonyl S atom of IPETC donates some electrons to the 3p orbital of S atom on elemental sulfur surface to form a normal covalent bond, while the 2p orbital of O atom of IPETC accepts the electrons from the 3p orbital of S atom in elemental sulfur surface to form a backdonation covalent bond.

Table 13. Mulliken charge populations of bonding atoms before and after adsorption.

Bonding Atom	Adsorption State	Electron Charge (e)			Charge Population
		s	p	d	
O $_{(O\text{-}S)}$	Before adsorption	1.77	4.66	0.00	−0.44
	After adsorption	1.77	4.71	0.00	−0.46
S $_{(O\text{-}S)}$	Before adsorption	1.89	4.19	0.00	−0.01
	After adsorption	1.89	4.12	0.00	−0.07
Carbonyl S $_{(S\text{-}S)}$	Before adsorption	1.82	4.41	0.00	−0.23
	After adsorption	1.82	4.10	0.00	0.05
S $_{(S\text{-}S)}$	Before adsorption	1.89	4.10	0.00	0.01
	After adsorption	1.89	4.12	0.00	−0.01

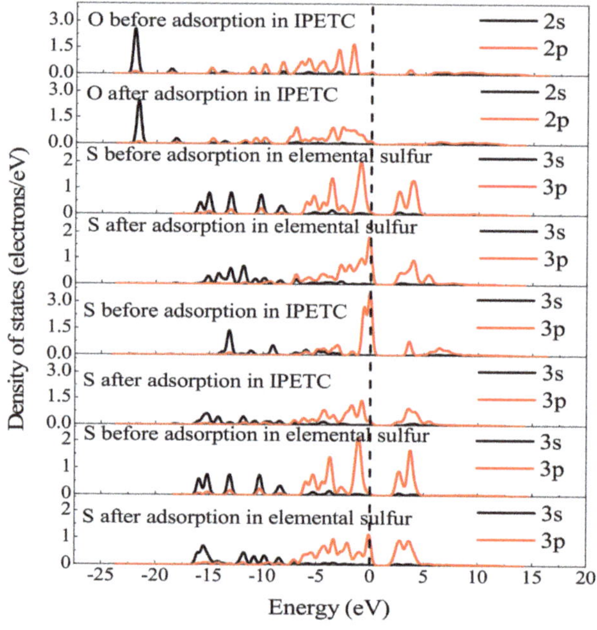

Figure 11. Densities of states of bonding atoms before and after IPETC adsorption on elemental sulfur (100) plane.

The 2p state of O atom of IPETC slightly shifts to the higher energy direction, which means that it gets some electrons in the reaction. However, there is no evident variation in the 3p state of bonding S atom on elemental sulfur surface, indicating that the interaction between O and S atoms is weak. The 3p state of carbonyl S atom of IPETC moves towards the lower energy direction, which indicates that when the S atom of IPETC reacts with an S atom on elemental sulfur surface, it loses some electrons and its oxidation is strengthened.

At the same time, the localization of 3p state of bonding S atom on elemental sulfur surface declines and evident hybridization happens on the 3p orbital of carbonyl S atom and 3p orbital of S atom from -5 eV to 0 eV. Thus, the interaction between carbonyl S atom in IPETC and S atom in elemental sulfur surface is strong.

Based on the above results, it can be concluded that when IPETC interacts with elemental sulfur surface, carbonyl S of IPETC offers some electrons to the S atom on mineral surface to form a normal covalent bond, while O of IPETC accepts some electrons from the S atom to generate a backdonation covalent bond. The backdonation covalent bond is relatively weak, and thus a normal covalent bond plays a leading role.

3.4.3. FTIR Spectrum Analysis

The infrared spectra of IPETC, elemental sulfur and IPETC-adsorbed elemental sulfur are presented in Figure 12. After interaction with IPETC, the infrared spectrum of elemental sulfur changed significantly, suggesting that IPETC chemisorbed on elemental sulfur surface. Two new strong absorption peaks occurred at 2318.44 cm^{-1} and 503.42 cm^{-1}, which are separately attributed to C=S-S and C-O-S coupled vibrations. The possible reason for this may be that when IPETC reacted with elemental sulfur, the S and O atoms of C=S and C-O bonds separately lost and obtained some electrons. As a result of this, a normal covalent bond and a backdonation covalent bond were formed between C=S and C-O bonds in IPETC and S atoms in elemental sulfur surface, resulting in the generation of two new coupled vibration peaks.

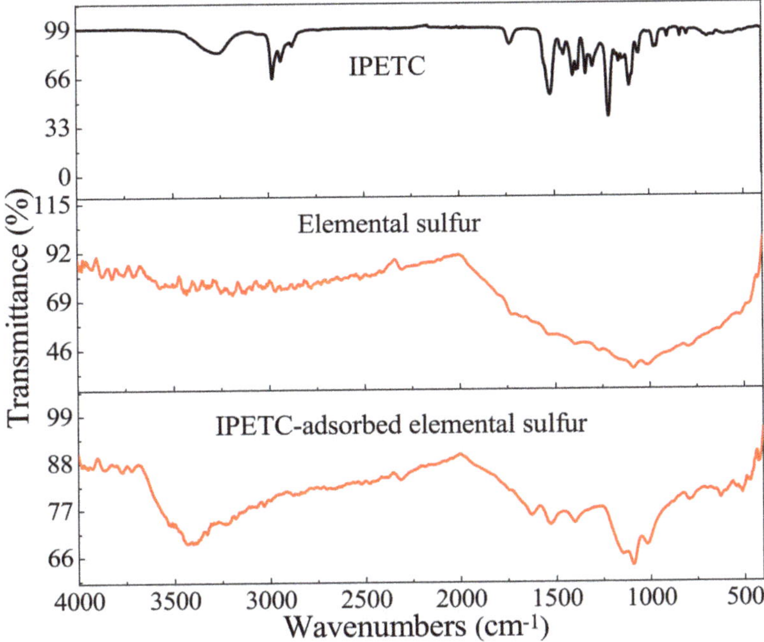

Figure 12. FTIR spectra of IPETC, elemental sulfur and IPETC-adsorbed elemental sulfur.

4. Conclusions

In this research, O-Isopropyl-N-Ethyl thionocarbamate (IPETC) collector was used to selectively recover elemental sulfur from a high-sulfur residue. The raw ore flotation test showed that in comparison with sodium ethyl xanthate (SEX) and ammonium dibutyl dithiophosphate (ADDTP) collectors, IPETC exhibited a superior collecting ability and selectivity to the elemental sulfur. Pure mineral flotation and adsorption experiments further proved that the collecting power and selectivity of IPETC to elemental sulfur are the optimum among the three collectors.

The bond length of C=S/P=S covalent bond, Mulliken population and frontier orbital energy difference between collectors and minerals manifested that the selectivity of three collectors to elemental sulfur is in the order of IPETC > ADDTP > SEX. The density of states and frontier orbital analyses showed that the HOMO of each collector is mainly constituted by electron orbitals of carbonyl S atom, the LUMOs of IPETC and SEX are comprised of conjugated big π-bond formed by electron orbitals of the coplanar functional groups, and the LUMO of ADDTP is composed of the electron orbitals of O, S and P atoms.

The adsorption configuration analysis indicated that the steadiest adsorption configuration of IPETC on the surfaces of elemental sulfur, chalcopyrite, sphalerite, pyrite, anglesite and albite is the simultaneous adsorption of carbonyl S together with O, and IPETC adsorption on elemental sulfur surface is the most stable. The Mulliken population and density of state analyses of bonding atoms indicated that when IPETC reacts with the elemental sulfur surface, the 3p orbital of carbonyl S atom of IPETC donates electrons to the 3p orbital of S atom in elemental sulfur surface to form normal covalent bond while the 2p orbital of O atom of IPETC obtains the electrons from the 3p orbital of S atom in elemental sulfur surface to generate a backdonation covalent bond, and the stronger normal covalent bond plays a dominant role. The FTIR spectrum analysis supported the generation of a normal covalent bond and a backdonation covalent bond during IPETC adsorption on elemental sulfur surface.

Author Contributions: Conceptualization, Z.D. and B.X.; methodology, G.L. and B.Z.; software, F.Z.; validation, T.J.; formal analysis, F.W.; investigation, G.L.; resources, G.L.; data curation, Z.D.; writing—original draft preparation, G.L. and Z.D.; writing—review and editing, Z.D. and B.X.; supervision, B.X. and T.J.; funding acquisition, G.L. All authors have read and agreed to the published version of the manuscript.

Funding: This research was funded by National Key Research and Development Program of China (Nos. 2018YFC1902005 and 2018YFC1902006).

Institutional Review Board Statement: Not applicable.

Informed Consent Statement: Not applicable.

Data Availability Statement: Data available in a publicly accessible repository.

References

1. Abbasi, A.; Nasef, M.M.; Yahya, W.Z.N. Sulfur based polymers by inverse vulcanization: A novel path to foster green chemistry. *Green Mater.* **2020**, *8*, 172–180. [CrossRef]
2. Dong, Z.L.; Jiang, T.; Xu, B.; Yang, Y.B.; Li, Q. Recovery of gold from pregnant thiosulfate solutions by the resin adsorption technique. *Metals* **2017**, *7*, 555. [CrossRef]
3. Dong, Z.L.; Jiang, T.; Xu, B.; Yang, Y.B.; Li, Q. An eco-friendly and efficient process of low potential thiosulfate leaching-resin adsorption recovery for extracting gold from a roasted gold concentrate. *J. Clean. Prod.* **2019**, *229*, 387–398. [CrossRef]
4. Dong, Z.L.; Jiang, T.; Xu, B.; Yang, J.K.; Chen, Y.Z.; Yang, Y.B.; Li, Q. Comprehensive recoveries of selenium, copper, gold, silver and lead from a copper anode slime with a clean and economical hydrometallurgical process. *Chem. Eng. J.* **2020**, *393*, 124762. [CrossRef]
5. Xu, B.; Kong, W.H.; Li, Q.; Yang, Y.B.; Jiang, T. A review of thiosulfate leaching of gold: Focus on thiosulfate consumption and gold recovery from pregnant solution. *Metals* **2017**, *7*, 222. [CrossRef]
6. Xu, B.; Li, K.; Dong, Z.L.; Yang, Y.B.; Li, Q.; Liu, X.L.; Jiang, T. Eco-friendly and economical gold extraction by nickel catalyzed ammoniacal thiosulfate leaching-resin adsorption recovery. *J. Clean. Prod.* **2019**, *233*, 1475–1485. [CrossRef]
7. Xu, B.; Chen, Y.Z.; Dong, Z.L.; Jiang, T.; Zhang, B.S.; Liu, G.Q.; Yang, J.K.; Li, Q.; Yang, Y.B. Eco-friendly and efficient extraction of valuable elements from copper anode mud using an integrated pyro-hydrometallurgical process. *Resour. Conserv. Recycl.* **2021**, *164*, 105195. [CrossRef]
8. Gu, Y.; Zhang, T.A.; Liu, Y.; Mu, W.Z.; Zhang, W.G.; Dou, Z.H.; Jiang, X.L. Pressure acid leaching of zinc sulfide concentrate. *Trans. Nonferr. Met. Soc. China* **2010**, *20*, s136–s140. [CrossRef]

9. Padilla, R.; Vega, D.; Ruiz, M.C. Pressure leaching of sulfidized chalcopyrite in sulfuric acid–oxygen media. *Hydrometallurgy* **2010**, *86*, 80–88. [CrossRef]

10. Jorjani, E.; Ghahreman, A. Challenges with elemental sulfur removal during the leaching of copper and zinc sulfides, and from the residues; a review. *Hydrometallurgy* **2017**, *171*, 333–343. [CrossRef]

11. Forward, F.A.; Veltman, H. Direct leaching zinc-sulfide concentrates by Sherritt Gordon. *JOM* **1959**, *11*, 836–840. [CrossRef]

12. Corriou, J.P.; Gély, R.; Viers, P. Thermodynamic and kinetic study of the pressure leaching of zinc sulfide in aqueous sulfuric acid. *Hydrometallurgy* **1988**, *21*, 85–102. [CrossRef]

13. Lampinen, M.; Laari, A.; Turunen, I. Kinetic model for direct leaching of zinc sulfide concentrates at high slurry and solute concentration. *Hydrometallurgy* **2015**, *153*, 160–169. [CrossRef]

14. Mu, W.Z.; Zhang, T.A.; Liu, Y.; Gu, Y.; Dou, Z.H.; Lv, G.Z.; Bao, L.; Zhang, W.G. E-pH diagram of ZnS-H2O system during high pressure leaching of zinc sulfide. *Hydrometallurgy* **2010**, *20*, 2012–2019. [CrossRef]

15. Liu, G.Q.; Jiang, K.X.; Zhang, B.S.; Dong, Z.L.; Zhang, F.; Wang, F.; Jiang, T.; Xu, B. Selective flotation of elemental sulfur from pressure acid leaching residue of zinc sulfide. *Minerals* **2021**, *11*, 89. [CrossRef]

16. Qin, S.C.; Jiang, K.X.; Wang, H.B.; Zhang, B.S.; Wang, Y.F.; Zhang, X.D. Research on behavior of iron in the zinc sulfide pressure leaching process. *Minerals* **2020**, *10*, 224. [CrossRef]

17. Rao, S.; Wang, D.X.; Liu, Z.Q.; Zhang, K.F.; Cao, H.Y.; Tao, J.Z. Selective extraction of zinc, gallium, and germanium from zinc refinery residue using two stage acid and alkaline leaching. *Hydrometallurgy* **2019**, *183*, 38–44. [CrossRef]

18. Wang, Z.Y.; Cai, X.L.; Zhang, Z.B.; Zhang, L.B.; Wang, S.X.; Peng, J.H. Separation and enrichment of elemental sulfur and mercury from hydrometallurgical zinc residue using sodium sulfide. *Trans. Nonferr. Met. Soc. China* **2015**, *25*, 640–646. [CrossRef]

19. Halfyard, J.E.; Hawboldt, K. Separation of elemental sulfur from hydrometallurgical residue: A review. *Hydrometallurgy* **2011**, *109*, 80–89. [CrossRef]

20. Li, H.L.; Yao, X.L.; Wang, M.X.; Wu, S.K.; Ma, W.W.; Wei, W.W.; Li, L.Q. Recovery of elemental sulfur from zinc concentrate direct leaching residue using atmospheric distillation: A pilot-scale experimental study. *J. Air Waste Manag.* **2014**, *64*, 95–103. [CrossRef]

21. Li, H.L.; Wu, X.Y.; Wang, M.X.; Wang, J.; Wu, S.K.; Yao, X.L.; Li, L.Q. Separation of elemental sulfur from zinc concentrate direct leaching residue by vacuum distillation. *Sep. Purif. Technol.* **2014**, *138*, 41–46. [CrossRef]

22. Chen, J.H.; Lan, L.H.; Chen, Y. Computational simulation of adsorption and thermodynamic study of xanthate, dithiophosphate and dithiocarbamate on galena and pyrite surfaces. *Miner. Eng.* **2013**, *46–47*, 136–143. [CrossRef]

23. Huang, Z.Q.; Zhong, H.; Wang, S.; Xia, L.Y.; Zou, W.B.; Liu, G.Y. Investigations on reverse cationic flotation of iron ore by using a Gemini surfactant: Ethane-1,2-bis (dimethyl-dodecyl-ammonium bromide). *Chem. Eng. J.* **2014**, *257*, 218–228. [CrossRef]

24. Ma, X.; Xia, L.Y.; Wang, S.; Zhong, H.; Jia, H. Structural modification of xanthate collectors to enhance the flotation selectivity of chalcopyrite. *Ind. Eng. Chem. Res.* **2017**, *56*, 6307–6316. [CrossRef]

25. Wang, Z.; Xu, L.H.; Wang, J.M.; Wang, L.; Xiao, J.H. A comparison study of adsorption of benzohydroxamic acid and amyl xanthate on smithsonite with dodecylamine as co-collector. *Appl. Surf. Sci.* **2017**, *426*, 1141–1147. [CrossRef]

26. Liu, F.P.; Wang, J.L.; Peng, C.; Liu, Z.H.; Wilson, B.P.; Lundström, M. Recovery and separation of silver and mercury from hazardous zinc refinery residues produced by zinc oxygen pressure leaching. *Hydrometallurgy* **2019**, *185*, 38–45. [CrossRef]

27. Fan, Y.Y.; Liu, Y.; Niu, L.P.; Jing, T.L.; Zhang, T.A. Separation and purification of elemental sulfur from sphalerite concentrate direct leaching residue by liquid paraffin. *Hydrometallurgy* **2019**, *186*, 162–169. [CrossRef]

28. Xing, P.; Ma, B.Z.; Wang, C.Y.; Wang, L.; Chen, Y.Q. A simple and effective process for recycling zinc-rich paint residue. *Waste Manag.* **2018**, *76*, 234–241. [CrossRef]

29. Qiu, T.S.; He, Y.Q.; Qiu, X.H.; Yang, X.L. Density functional theory and experimental studies of Cu^{2+} activation on a cyanide-leached sphalerite surface. *J. Ind. Eng. Chem.* **2017**, *45*, 307–315. [CrossRef]

30. Rashchi, F.; Dashti, A.; Arabpour-Yazdi, M.; Abdizadeh, H. Anglesite flotation: A study for lead recovery from zinc leach residue. *Miner. Eng.* **2005**, *18*, 205–212. [CrossRef]

31. Dong, Z.L.; Jiang, T.; Xu, B.; Li, Q.; Zhong, H.; Yang, Y.B. Selective flotation of galena using a novel collector S-benzyl-N-ethoxycarbonyl thiocarbamate: An experimental and theoretical investigation. *J. Mol. Liq.* **2021**, *330*, 115643. [CrossRef]

32. Xu, B.; Wu, J.T.; Dong, Z.L.; Jiang, T.; Li, Q.; Yang, Y.B. Flotation performance, structure-activity relationship and adsorption mechanism of a newly-synthesized collector for copper sulfide minerals in Gacun polymetallic ore. *Appl. Surf. Sci.* **2021**, *551*, 149420. [CrossRef]

33. Le, M.N.; Lee, M.S. Hydrometallurgical treatment of elemental sulfur in spent catalysts by aqueous and nonaqueous solutions at low temperature. *Miner. Process. Extr. Metall. Rev.* **2020**, *41*, 217–226. [CrossRef]

34. Jia, Y.; Huang, K.H.; Wang, S.; Cao, Z.F.; Zhong, H. The selective flotation behavior and adsorption mechanism of thiohexanamide to chalcopyrite. *Miner. Eng.* **2019**, *137*, 187–199. [CrossRef]

Article

Hydrochloric Acid Leaching Behaviors of Copper and Antimony in Speiss Obtained from Top Submerged Lance Furnace

Sujin Chae [1], Kyoungkeun Yoo [1,*], Carlito Baltazar Tabelin [2] and Richard Diaz Alorro [3]

[1]   Department of Energy & Resources Engineering, Korea Maritime and Ocean University (KMOU), 727, Taejong-ro, Yeongdo-gu, Busan 49112, Korea; sujin.chae.082@gmail.com

[2]   School of Minerals and Energy Resources Engineering, University of New South Wales, Sydney 2052, NSW, Australia; c.tabelin@unsw.edu.au

[3]   Western Australian School of Mines: Minerals, Energy and Chemical Engineering, Faculty of Science and Engineering, Curtin University, Kalgoorlie 6430, WA, Australia; richard.alorro@curtin.edu.au

*   Correspondence: kyoo@kmou.ac.kr

Received: 28 September 2020; Accepted: 16 October 2020; Published: 20 October 2020

Abstract: Copper (Cu) has been recovered from speiss generated from top submerged lance furnace process, but it was reported that the leaching efficiency of Cu in sulfuric acid solution decreased with increasing antimony (Sb) content in the speiss. Scanning electron microscopy (SEM)–energy-dispersive X-ray spectroscopy (EDS) results indicate that Sb exists as CuSb alloy, which would retard the leaching of Cu. Therefore, hydrochloric acid leaching with aeration was performed to investigate the leaching behaviors of copper and antimony. The leaching efficiency of Cu increased with increasing agitation speed, temperature, HCl concentration, and the introduction ratio of O_2, but also with decreasing pulp density. The leaching efficiency of Cu increased to more than 99% within 60 min in 1 mol/L HCl solution at 600 rpm and 90 °C with 10 g/L pulp density and 1000 cc/min O_2. The leaching efficiency of Sb increased and then decreased in all 1 mol/L HCl leaching tests, and precipitate was observed in the leach solution, which was determined to be SbOCl or Sb_2O_3 by XRD analyses. However, in 2 mol/L–5 mol/L HCl solutions, the leaching efficiency of Sb increased to more than 95% (about 900 mg/L) and remained, so more than 2 mol/L HCl could stabilize Sb ion in the HCl solution.

Keywords: copper; antimony; hydrochloric acid leaching; speiss

1. Introduction

Top submerged lance (TSL) technology has been used to recover valuable metals from the by-product of zinc smelting or industrial waste [1–3]. The valuable metals are concentrated in speiss, and impurities such as Fe are discarded as slag [4]. Although the composition of the speiss generated in the TSL process varies depending on input materials, generally, in South Korea, the speiss contains Cu, Sb, and precious metals as the main components. The recovery process of valuable metals from the speiss is summarized as shown Figure 1. From the speiss, Cu component is leached with sulfuric acid, and so the precious metals are concentrated in leach residue. Other impurities, such as Sb and Pb, are discarded pyrometallurgically, and Ag and Au components are recovered by electrorefining processes in turn.

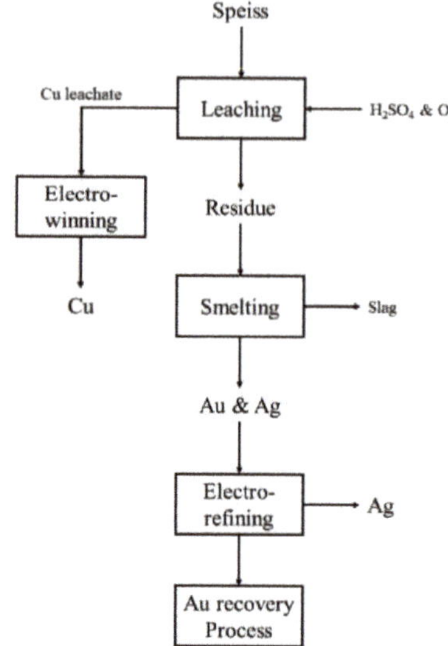

Figure 1. Schematic diagram of metal recovery process from speiss of top submerged lance process.

The high leaching efficiency of Cu from the speiss was required during the sulfuric acid leaching process because Cu in the residue reduces the purity of precious metal during the Au and Ag recovery process. Recently, as various secondary resources have been added to the TSL process, it was observed that the amount of antimony (Sb) increased and the leaching efficiency of Cu decreased. These facts, as a result, reduced the recovery efficiency of Au and Ag from subsequent processes. Therefore, although the leaching process of the speiss containing Cu and Sb should be improved, the leaching behaviors of materials containing Cu and Sb have been rarely reported.

The standard reduction potential of Cu is found to be 0.34 V [5,6], which indicates that Cu metal cannot be oxidized and dissolved by sulfuric acid. Additional oxidants such as Fe^{3+} [7,8], O_2 [9], Cl_2 [10] and H_2O_2 [11] or leaching media such as HNO_3 [12,13], HCl [14–17], and NH_3 [18,19] have been used to leach Cu metals. The leaching of Sb in sulfuric acid solution has rarely been investigated, and the solubility of Sb_2O_3 or Sb_2O_5, which could be generated from the oxidation of Sb, was found to be low [20]. The leaching of Sb has been investigated in sulfides using alkaline solution with Na_2S [21,22] and HCl with ozone [23], and from Pb dross [24] or fly ash [25] using hydrochloric acid. However, there have been few reports regarding Sb leaching, and the effect of Sb on the Cu leaching has not been reported.

Therefore, in the present study, the leach residue of the speiss was examined with SEM-EDS to understand the effect of Sb on the sulfuric acid leaching of the speiss, and then hydrochloric acid leaching tests with oxygen were performed to enhance the Cu leaching efficiency from the speiss. The effects of leaching factors such as temperature, agitation speed, HCl concentration, pulp density, and the flow rate of O_2 on the leaching behaviors of Cu and Sb were investigated in hydrochloric acid solution.

2. Materials and Methods

The speiss samples used in this study were obtained from a zinc smelter in Korea, and the D_{50} and D_{90} of the speiss samples were 307.4 μm and 795.5 μm, respectively. The speiss contained 77.16% Cu, 8.82% Sb, and 5.15% Pb as main components. Sulfuric acid (H_2SO_4, 95%, Junsei Co., Ltd.: Tokyo, Japan) and hydrochloric acid (HCl, 35%, Junsei Co., Ltd.: Tokyo, Japan) were used as leaching media, and purities of O_2, and N_2 were 99.99% and 99.5%, respectively.

Leaching tests of the speiss were performed in a 500 mL five-necked Pyrex glass reactor using a heating mantle to maintain temperature. The reactor was fitted with an agitator and a reflux condenser, which was used to prevent solution loss at high temperatures. In a typical run, 200 mL of acid solution (2 mol/L H_2SO_4 or 0.1–5 mol/L HCl) was poured into the reactor and, after the temperature of solution reached the thermal equilibrium (30–90 °C) at 200–800 rpm agitation speed, 2 g of the speiss sample was added to the reactor in the experiments except the pulp density test. During the tests, gases such as O_2, air, and N_2 were introduced at 200 cc/min–1000 cc/min, and 3 cm^3 of the solution was withdrawn periodically at desired time intervals (5–360 min) with a syringe. The samples were filtered with a 0.22 μm membrane filter and then the filtrate was diluted with 2% HNO_3 solution for Cu analysis and 3 mol/L HCl solution for Sb measurement, respectively.

The concentration of Cu and Sb in the solutions was measured with an inductively coupled plasma-atomic emission spectrometry (ICP-OES, Optima-8300, Perkin Elmer Inc.: Waltham, MA, USA). The precipitate generated during the leaching test was filtered and then dried at 105 °C. The precipitate was analyzed with an X-ray diffractometer (XRD, Smartlab, Rigaku Co.: Tokyo, Japan). The leaching residue obtained from the sulfuric acid leaching test was polished and then analyzed with a Scanning Electron Microscope (SEM, MIRA-3, Tescan Co.: Brno, Czech).

3. Results and Discussion

After sulfuric acid leaching of the speiss had been performed in 2 mol/L sulfuric acid solution at 400 rpm and 90 °C with 1% pulp density and 1000 cc/min introduction of O_2, the leaching residue was obtained and then examined with SEM-EDS. Figure 2 shows the SEM image of the leaching residue, and the center and outer parts of the particle shows different shapes. A net shape was observed in the entire cross section of the particle, and the net shape was found to be a CuSb intermetallic alloy, based on the result of SEM-EDS analysis. In the center of the particle, the net shape is full of dark parts, which were found to be Cu, while only the net shape was observed, without Cu, in the outer part of the particle. These results indicate that Cu was leached from the outer of particle, but Cu remained inside the CuSb alloy net shape in the center of particle because CuSb would prevent leaching of Cu. It has been reported that the leaching efficiency of Sb is lower or slower than Cu [26], and that the solubility of Sb is low in sulfuric acid [20]. Therefore, Sb should be dissolved to enhance the leaching efficiency of Cu, so hydrochloric acid leaching tests were performed because chloride ion could make complex ions with metal, thus enhancing the solubility of the metal [27].

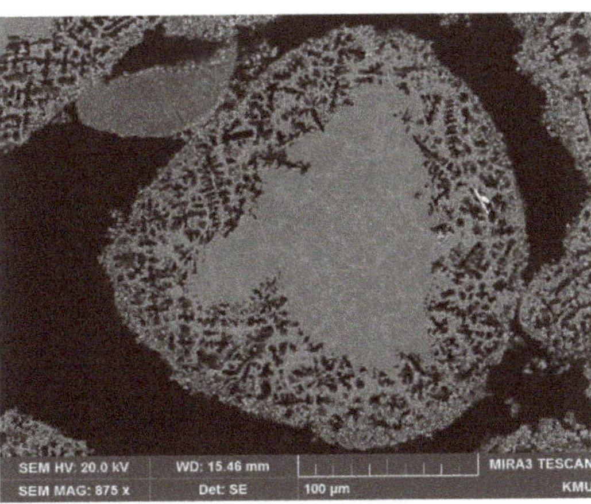

Figure 2. SEM image of sulfuric acid leaching residue of the speiss.

Leaching tests of the speiss at agitation speeds in the range of 200–800 rpm were performed to investigate the effect of the liquid film boundary diffusion surrounding the particles on the leaching efficiency of Cu and Sb in 1 mol/L HCl at 90 °C with 1% pulp density and 1000 cc/min O_2. As shown in Figure 3a, the leaching efficiency of Cu increased rapidly, and then gradually. The efficiency increased when the agitation speed was increased from 200 rpm to 600 rpm, while the efficiencies show similar leaching behaviors at 600 rpm and 800 rpm, so working agitation speed was fixed at 600 rpm to ensure effective particle suspension in the solution in all the subsequent leaching tests. Figure 3b shows the leaching behaviors of Sb, where the leaching efficiency of Sb increased to more than 99% and then decreased with time. The efficiency at 200 rpm decreased more slowly than those between 400 rpm and 800 rpm, and, at 360 min, lower leaching efficiency of Sb was observed at higher agitation speeds. When the concentration of Sb began to decrease, precipitate was observed in the solution. Figure 4 shows the XRD pattern of the precipitate, which was determined to be antimony oxychloride (SbOCl). These results indicate that Cu was leached from the speiss by HCl leaching, while Sb dissolved and was then precipitated as SbOCl.

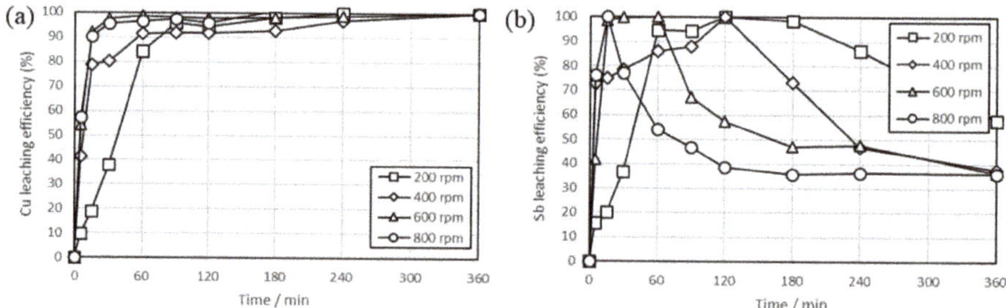

Figure 3. Leaching efficiencies of (**a**) Cu and (**b**) Sb with time in the function of agitation speed.

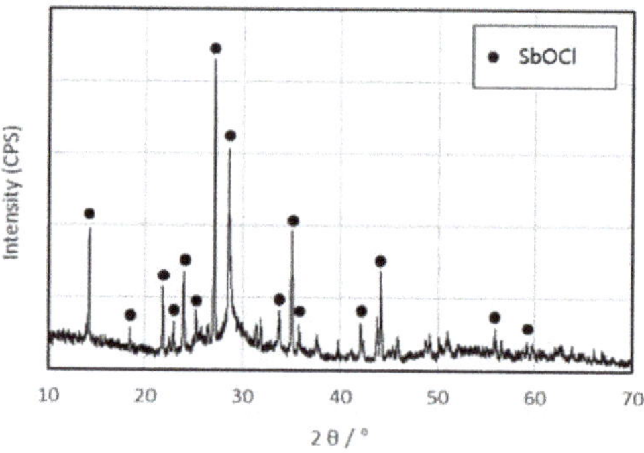

Figure 4. XRD pattern of leaching residue obtained from HCl leaching test in Figure 3.

In the TSL process, the metal product was generated as elemental metal or intermetallic alloy, as shown in Figure 1. The leaching of Cu metal can be summarized using the following equation [28]:

$$Cu^{2+} + Cu = 2Cu^+,\tag{1}$$

where cupric ion (Cu^{2+}) could oxidize Cu metal in hydrochloric acid or ammonia solution [18,19,28] into cuprous ion (Cu^+). This Cu^+ ion is unstable, an so is easily oxidized into Cu^{2+} as follows [6]:

$$2Cu^+ + 2H^+ + 1/2O_2 = 2\,Cu^{2+} + H_2O\tag{2}$$

The regenerated Cu^{2+} could be reused as an oxidant for Cu metal leaching. Therefore, it is important to introduce oxygen into the leaching solution for Cu metal leaching. The leaching tests of the speiss were performed to investigate the effect of gas introduction on the leaching efficiency of Cu and Sb in 1 mol/L HCl at 90 °C and 600 rpm with 1% pulp density. The O_2 introduction rate was adjusted from 200 cc/min to 1000 cc/min, and 1000 cc/min air and N_2 were also introduced into the solution for comparison. As shown in Figure 5a, the leaching efficiency of Cu increased to more than 99% within 60 min in the cases of O_2 introduction, whereas the efficiency with air introduction increased rapidly and then gradually to 99% within 360 min. When N_2 was introduced, the efficiency increased gradually to 18% within 360 min. These results indicate that the anoxic condition prevents leaching of Cu, because the regeneration of Cu^{2+}, as shown in Equation (2), is suppressed, although 18% of Cu dissolved under the N_2 introduction, which is due to partly oxidized surface of Cu [9]. Therefore, O_2 should be introduced into the leach solution for Cu metal leaching, but the leaching efficiency of Cu increased regardless of the O_2 introduction rate. Figure 5b shows that the leaching efficiency of Sb increased and then decreased with time except with the introduction of N_2, where the leaching efficiency of Sb was very low. An introduction rate of 1000 cc/min of O_2 could retard the decrease in Sb concentration, so the gas introduction rate was fixed at 1000 cc/min O_2 in all subsequent leaching tests.

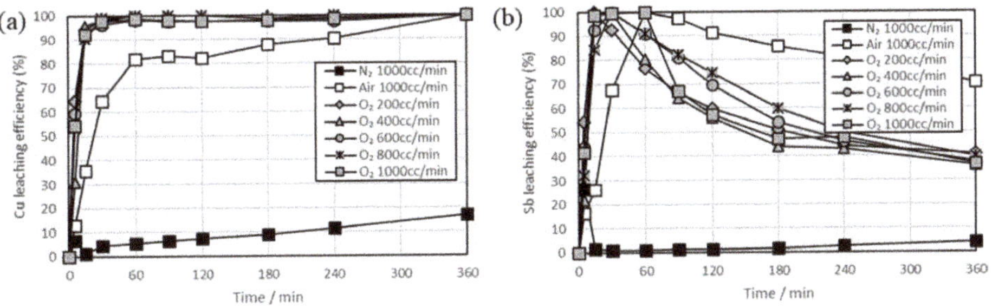

Figure 5. Leaching efficiencies of (**a**) Cu and (**b**) Sb with time in the function of gas introduction rate.

The effects of HCl concentration on the leaching efficiency of Cu and Sb were investigated under the following leaching conditions: 600 rpm at 90 °C with 1% pulp density and 1000 cc/min O_2. In the case of Cu leaching, as shown in Figure 6a, higher leaching efficiency of Cu was observed at higher HCl concentrations at 15 min, and then similar leaching behaviors of Cu were observed except for the test in 0.1 mol/L HCl. In Figure 6b, low leaching efficiency of Sb was observed in 0.1 mol/L HCl solution over the entire leaching time, and the efficiency increased and then decreased with time in the 1 mol/L HCl solution. When HCl concentration was increased to more than 2 mol/L, the leaching efficiency of Sb increased to more than 95% and then remained there. The concentration of Sb was measured to be about 900 mg/L. These results indicate that more than 2 mol/L HCl could stabilize Sb ions in the solution with the Sb concentration range, and this fact has rarely been reported in conventional studies.

The leaching tests of the speiss were performed to investigate the effect of temperature on the leaching efficiency of Cu and Sb in 1 mol/L HCl at 30–90 °C and 600 rpm with 1% pulp density and 1000 cc/min O_2. Figure 7a shows that the leaching efficiency of Cu increased more rapidly at higher temperature, and that the efficiency increased to more than 99% within 120 min at 50–90 °C and within 360 min at 30 °C. In the case of Sb leaching as shown in Figure 7b, the leaching efficiency of Sb increased more rapidly at higher temperature during the early leaching time, except the test at 30 °C, and the efficiency decreased gradually. Although the leaching efficiency of Sb at 30 °C increased to 53% at 60 min, that of Cu increased as shown in Figure 7a. Since the precipitate of Sb was observed in the solution during the leaching test at 30 °C, Cu continued to dissolve while Sb dissolved and then precipitated.

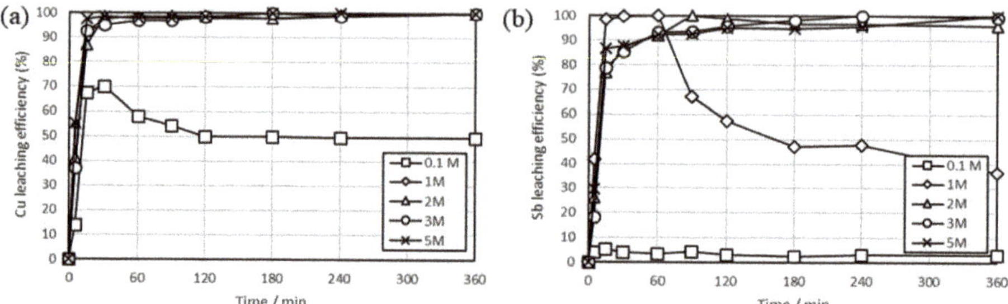

Figure 6. Leaching efficiencies of (**a**) Cu and (**b**) Sb with time in the function of HCl concentration.

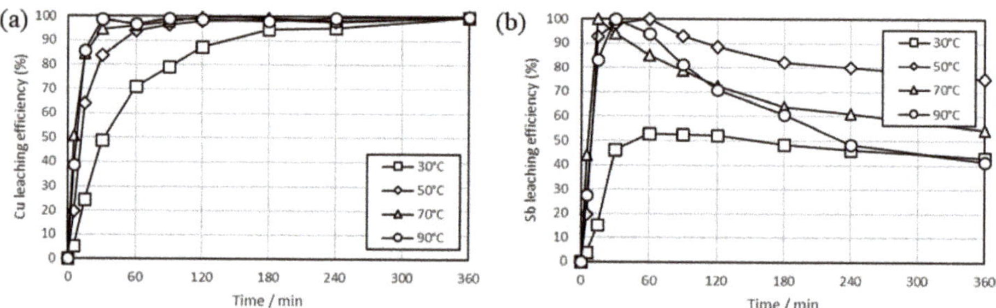

Figure 7. Leaching efficiencies of (**a**) Cu and (**b**) Sb with time as a function of temperature.

The effects of pulp density on the leaching efficiency of Cu and Sb were investigated in 1 mol/L HCl solution at 90 °C and 600 rpm with 1000 cc/min O_2. As shown in Figure 8a, the leaching efficiency of Cu decreased when increasing pulp density from 10 g/L to 50 g/L. In the case of Sb leaching, except the pulp density of 10 g/L, Figure 8b shows that the concentration of Sb increased and then decreased rapidly to about 10%, and that the concentration of Sb in the leaching test with 10 g/L increased and then decrease gradually to 40%. The precipitates formed in the leaching test with 20 g/L to 50 g/L were collected and analyzed with XRD. As shown in Figure 9, $CuCl_2 \cdot 3Cu(OH)_2$ was observed in the precipitate formed in the test with 50 g/L, and, in the test with 20 g/L or 30 g/L, the precipitate is determined to be Sb_2O_3. The pHs of the leaching solution increased to 3.8 and 4.2 in the leaching tests with 20 g/L and 50 g/L, respectively. Therefore, the Cu precipitate in the test with 50 g/L resulted from the higher pH and pulp density.

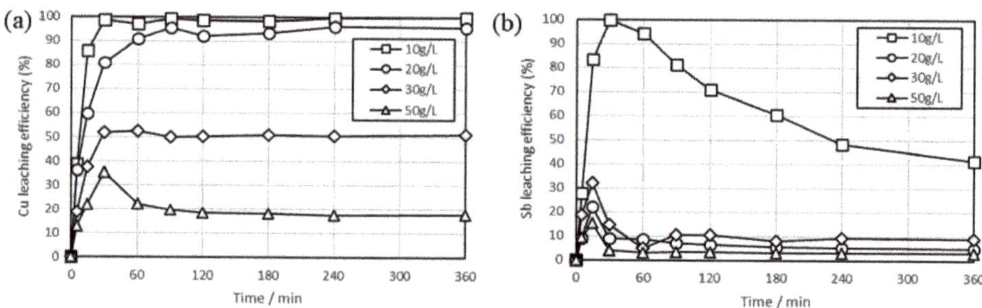

Figure 8. Leaching efficiencies of (**a**) Cu and (**b**) Sb with time in the function of pulp density.

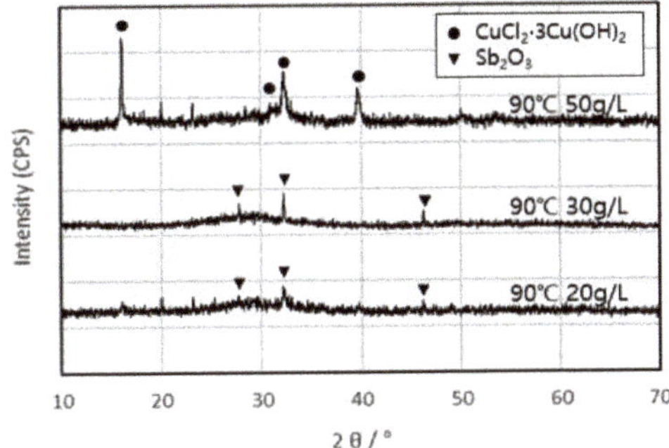

Figure 9. XRD pattern of leaching residue obtained from HCl leaching test in Figure 8.

4. Conclusions

SEM-EDS analysis of the leaching residue obtained from H_2SO_4 leaching of speiss and HCl leaching of speiss was performed to enhance the leaching efficiency of Cu and to investigate the leaching behavior of Sb in HCl solution.

The SEM-EDS results showed that CuSb alloy remained in the leaching residue even after Cu was leached by H_2SO_4 leaching, which indicates that the leaching rate of CuSb is slower than Cu, and that CuSb could retard the leaching of Cu. Therefore, HCl was used as a leaching agent to leach the speiss. The leaching efficiency of Cu increased with increasing agitation speed, temperature, HCl concentration, gas introduction rate but with decreasing pulp density. More than 99% of Cu was dissolved for 60 min in 1 mol/L HCl solution at 600 rpm and 90 °C with 10 g/L pulp density and 1000 cc/min O_2. The leaching efficiency of Sb increased, and then was precipitated in all 1 mol/L HCl leaching tests, and the precipitate was determined to be SbOCl. However, since more than 95% of Sb (about 900 mg/L) was dissolved and remained in 2 mol/L–5 mol/L HCl solutions, this result indicates that 2 mol/L or more concentrated HCl could stabilize Sb ion in the HCl solution after 60 min.

Author Contributions: Methodology, S.C. and K.Y.; writing—original draft preparation, S.C., C.B.T. and K.Y.; project administration and funding acquisition, K.Y.; data curation, S.C. and K.Y.; writing—review and editing, providing ideas, R.D.A. and K.Y. All authors have read and agreed to the published version of the manuscript.

Funding: This work was supported by the Technology Innovation Program (or Industrial Strategic Technology Development Program—Development of Material Component Technology) (20011176, Development of advanced technology in Hydrometallurgy for high added value of resources recovery) funded By the Ministry of Trade, Industry & Energy (MOTIE, Korea).

Conflicts of Interest: The authors declare no conflict of interest.

References

1. Hoang, J.; Reuter, M.A.; Matusewicz, R.; Hughes, S.; Piret, N. Top submerged lance direct zinc smelting. *Miner. Eng.* **2009**, *22*, 742–751. [CrossRef]
2. Kim, B.S.; Jeong, S.B.; Lee, J.C.; Shin, D.; Moon, N.I. Behaviors of lead and zinc in top submerged lance (TSL) plant at Sukpo zinc refinery. *Mater. Trans.* **2012**, *53*, 985–990. [CrossRef]
3. Sohn, H.S. Current Status of Zinc Smelting and Recycling. *J. Korean Inst. Resour. Recycl.* **2019**, *28*, 30–41. [CrossRef]
4. Yoo, J.K.; Lee, M.; Kim, G.H.; Yoo, K. The Gravity Separation of Speiss and Limestone Granules Using Vibrating Zirconia Ball Bed. *J. Korean Inst. Resour. Recycl.* **2020**, *29*, 36–42. [CrossRef]
5. Brett, C.M.A.; Brett, A.M.O. *Electrochemistry*; Oxford University Press Inc.: New York, NY, USA, 1993; pp. 416–419.
6. Yoo, K.; Park, Y.; Choi, S.; Park, I. Improvement of Copper Metal Leaching in Sulfuric Acid Solution by Simultaneous Use of Oxygen and Cupric Ions. *Metals* **2020**, *10*, 721. [CrossRef]

7.   Lee, S.; Yoo, K.; Jha, M.K.; Lee, J. Separation of Sn from waste Pb-free Sn-Ag-Cu solder in hydrochloric acid solution with ferric chloride. *Hydrometallurgy* **2015**, *157*, 184–187. [CrossRef]

8.   Xing, W.; Lee, S.; Lee, M. Solvent extraction of Li(I) from weak HCl solution with the mixture of neutral extractants containing FeCl$_3$. *J. Korean Inst. Resour. Recycl.* **2018**, *27*, 53–58. [CrossRef]

9.   Park, I.; Yoo, K.; Alorro, R.D.; Kim, M.; Kim, S. Leaching of copper from cuprous oxide in aerated sulfuric acid. *Mater. Trans.* **2017**, *58*, 1500–1504. [CrossRef]

10.   Kim, E.Y.; Kim, M.S.; Lee, J.C.; Yoo, K.; Jeong, J. Leaching behavior of copper using electro-generated chlorine in hydrochloric acid solution. *Hydrometallurgy* **2010**, *100*, 95–102. [CrossRef]

11.   Kim, S.; Lee, J.C.; Lee, K.S.; Yoo, K.; Alorro, R.D. Separation of tin, silver and copper from waste Pb-free solder using hydrochloric acid leaching with hydrogen peroxide. *Mater. Trans.* **2014**, *55*, 1885–1889. [CrossRef]

12.   Yoo, K.; Lee, J.C.; Lee, K.S.; Kim, B.S.; Kim, M.S.; Kim, S.K.; Pandey, B.D. Recovery of Sn, Ag and Cu from waste Pb-free solder using nitric acid leaching. *Mater. Trans.* **2012**, *53*, 2175–2180. [CrossRef]

13.   Yoo, K.; Lee, K.; Jha, M.K.; Lee, J.; Cho, K. Preparation of nano-sized tin oxide powder from waste Pb-free solder by direct nitric acid leaching. *J. Nanosci. Nanotechnol.* **2016**, *16*, 11238–11241. [CrossRef]

14.   Moon, G.; Yoo, K. Separation of Cu, Sn, Pb from photovoltaic ribbon by hydrochloric acid leaching with stannic ion followed by solvent extraction. *Hydrometallurgy* **2017**, *171*, 123–127. [CrossRef]

15.   Chen, W.S.; Chen, Y.J.; Yueh, K.C. Separation of Valuable Metal from Waste Photovoltaic Ribbon through Extraction and Precipitation. *J. Korean Inst. Resour. Recycl.* **2020**, *29*, 69–77. [CrossRef]

16.   Kang, Y.H.; Hyun, S.K. Study on the Preparation of Copper Sulfate by Copper Powder using Cation Membrane Electrowinning Prepared from Waste Cupric Chloride Solution. *J. Korean Inst. Resour. Recycl.* **2019**, *28*, 62–72. [CrossRef]

17.   Kim, S.K.; Lee, J.C.; Yoo, K. Leaching of tin from waste Pb-free solder in hydrochloric acid solution with stannic chloride. *Hydrometallurgy* **2016**, *165*, 143–147. [CrossRef]

18.   Jeon, S.; Yoo, K.; Alorro, R.D. Separation of Sn, Bi, Cu from Pb-free solder paste by ammonia leaching followed by hydrochloric acid leaching. *Hydrometallurgy* **2017**, *169*, 26–30. [CrossRef]

19.   Lim, Y.; Kwon, O.; Lee, J.; Yoo, K. The ammonia leaching of alloy produced from waste printed circuit boards smelting process. *Geosyst. Eng.* **2013**, *16*, 216–224. [CrossRef]

20.   Karlsson, T.Y. Studies on the Recovery of Secondary Antimony Compounds from Waste. Ph.D. Thesis, Chalmers University of Technology, Göteborg, Sweden, 2017; pp. 5–7.

21.   Awe, S.A.; Sundkvist, J.E.; Bolin, N.J.; Sandström, Å. Process flowsheet development for recovering antimony from Sb-bearing copper concentrates. *Miner. Eng.* **2013**, *49*, 45–53. [CrossRef]

22.   Awe, S.A.; Sandström, Å. Selective leaching of arsenic and antimony from a tetrahedrite rich complex sulphide concentrate using alkaline sulphide solution. *Miner. Eng.* **2010**, *23*, 1227–1236. [CrossRef]

23.   Guo, X.Y.; Xin, Y.T.; Wang, H.; Tian, Q. Leaching kinetics of antimony-bearing complex sulfides ore in hydrochloric acid solution with ozone. *Trans. Nonferrous Met. Soc. China* **2017**, *27*, 2073–2081. [CrossRef]

24.   Choi, S.; Yoo, K.; Alorro, R.D. Hydrochloric acid leaching behavior of metals from non-magnetic fraction of Pb dross. *Geosyst. Eng.* **2019**, *22*, 347–354. [CrossRef]

25.   Elomaa, H.; Seisko, S.; Lehtola, J.; Lundström, M. A study on selective leaching of heavy metals vs. iron from fly ash. *J. Mater. Cycles Waste Manag.* **2019**, *21*, 1004–1013. [CrossRef]

26.   Bae, E.; Yoo, K. Leaching behavior of valuable metals from by-product generated during purification of zinc electrolyte. *Geosyst. Eng.* **2016**, *19*, 312–316. [CrossRef]

27.   Oh, J.; Yoo, K.; Bae, M.; Kim, S.; Alorro, R.D. The adsorption behaviors of gold ions in simulated leachate using magnetite. *J. Korean Soc. Miner. Energy Resour. Eng.* **2019**, *56*, 79–85. [CrossRef]

28.   Lee, S.; Yoo, K.; Lee, J. Preparation of Cu$_2$O Powder in NaOH solution Using CuCl Obtained from Spent Printed Circuit Boards Etchant. *J. Korean Soc. Miner. Energy Resour. Eng.* **2018**, *55*, 194–199. [CrossRef]

Publisher's Note: MDPI stays neutral with regard to jurisdictional claims in published maps and institutional affiliations.

metals

Article

Copper Recovery and Reduction of Environmental Loading from Mine Tailings by High-Pressure Leaching and SX-EW Process

Labone L. Godirilwe [1], Kazutoshi Haga [1,*], Batnasan Altansukh [1], Yasushi Takasaki [1], Daizo Ishiyama [1], Vanja Trifunovic [2], Ljiljana Avramovic [2], Radojka Jonovic [2], Zoran Stevanovic [2] and Atsushi Shibayama [1,*]

[1] Department of Earth Resource Engineering and Environmental Science, Akita University, Akita 010-0865, Japan; llgodirilwe@gmail.com (L.L.G.); altansukh@gipc.akita-u.ac.jp (B.A.); yas-tksk@gipc.akita-u.ac.jp (Y.T.); ishiyama@gipc.akita-u.ac.jp (D.I.)
[2] Mining and Metallurgy Institute Bor, 19210 Bor, Serbia; vanja.trifunovic@irmbor.co.rs (V.T.); ljiljana.avramovic@irmbor.co.rs (L.A.); rafinacija@irmbor.co.rs (R.J.); zoran.stevanovic@irmbor.co.rs (Z.S.)
* Correspondence: khaga@gipc.akita-u.ac.jp (K.H.); sibayama@gipc.akita-u.ac.jp (A.S.)

Abstract: The flotation tailings obtained from Bor Copper Mine contain pyrite (FeS_2) and chalcopyrite ($CuFeS_2$), these sulfide minerals are known to promote acid mine drainage (AMD) which poses a serious threat to the environment and human health. This study focuses on the treatment of mine tailings to convert the AMD supporting minerals to more stable forms, while simultaneously valorizing the mine tailings. A combination of hydrometallurgical processes of high-pressure oxidative leaching (HPOL), solvent extraction (SX), and electrowinning (EW) were utilized to recover copper from mine tailings which contain about 0.3% Cu content. The HPOL process yielded a high copper leaching rate of 94.4% when water was used as a leaching medium. The copper leaching kinetics were promoted by the generation of sulfuric acid due to pyrite oxidation. It was also confirmed that a low iron concentration (1.4 g/L) and a high copper concentration (44.8 g/L) obtained in the stripped solution resulted in an improved copper electrodeposition current efficiency during copper electrowinning. Moreover, pyrite, which is primarily in the mine tailings, was converted into hematite after HPOL. A stability evaluation of the solid residue confirmed almost no elution of metal ions, confirming the reduced environmental loading of mine tailings through re-processing.

Keywords: tailings valorization; high-pressure leaching; copper recovery; acid mine drainage; metal elution

Citation: Godirilwe, L.L.; Haga, K.; Altansukh, B.; Takasaki, Y.; Ishiyama, D.; Trifunovic, V.; Avramovic, L.; Jonovic, R.; Stevanovic, Z.; Shibayama, A. Copper Recovery and Reduction of Environmental Loading from Mine Tailings by High-Pressure Leaching and SX-EW Process. *Metals* **2021**, *11*, 1335. https://doi.org/10.3390/met11091335

Academic Editors: Anna H. Kaksone and Chris Aldrich

Received: 16 July 2021
Accepted: 21 August 2021
Published: 24 August 2021

Publisher's Note: MDPI stays neutral with regard to jurisdictional claims in published maps and institutional affiliations.

1. Introduction

Mine tailings are waste material generated from the flotation process and there are billions of tons of already existing tailings all over the world. Moreover, several billion tons of additional mine tailings will inevitably be produced, as lower-grade and complex ores are being mined to sustain the world's growing demand for mineral resources [1,2]. Schlesinger et al. [3] (p.68) estimated that flotation tailings account for 98% of the ore fed in to concentrators, which are stored in large tailings dams near their mines. Unfortunately, the storage of mine tailings is associated with severe environmental challenges. If not managed properly, the failure of mine tailings storage facilities can result in catastrophic ramifications and environmental pollution, such as the well-known Brumadinho and Mariana tailings dam disasters in Brazil. The most serious environmental impact resulting from mine tailings is the generation of acid mine drainage (AMD). Mine tailings that contain metal sulfides, such as pyrite (FeS_2), are a major source of AMD [4–6]. Pyrite is a gangue mineral that is commonly found in mine tailings, when exposed to an environment with oxygen and water, oxidation of pyrite is promoted via reaction 1, which results in low pH conditions of less than 3 [5,7–9].

$$2FeS_2 + 7.5O_2 + H_2O \rightarrow Fe_2(SO_4)_3 + H_2SO_4 \tag{1}$$

The low pH conditions are detrimental because they increase the dissolution and mobility of toxic and heavy metals/metalloids such as Cd, Co, Cr, Cu, Fe, Mn, Ni, Pb, Zn, As, and Se [6,10]. When these toxic metals accumulate in the environment, they become a major source of contamination to water sources and potentially endanger ecosystems and human health [5,6,8,11]. The flotation tailings from Bor Copper Mine contain pyrite and chalcopyrite, therefore, management of these mine tailings is extremely important, as well as developing effective long-term strategies to reduce the environmental footprint of the mining industry.

The reprocessing of mine tailings is now being globally explored as an approach to be sustainable in both active and inactive mining operations. As new technologies emerge, valuable metals can efficiently be extracted from mine tailings with reduced environmental impacts. Most old copper mine tailings have a comparable copper grade to that of low-grade ores (0.2–0.3%) and considering their vast amount, they present a potential source of secondary raw material for copper production [6,7,12–15]. Several technologies have been used for the treatment of mine tailings to extract valuable metals, these mostly include bioleaching [12,15–17], flotation [10,14,18], hydrometallurgical methods [4,6,7,19], or a combination of these treatment methods [13,20]. Hydrometallurgical methods are commonly considered for the treatment of low-grade sulfide material because of comprehensive metal recovery and shorter retention times. Antonijevi'c et al. [7] investigated the leaching of flotation tailings from Bor Copper Mine using sulfuric acid and iron (III) sulfate as an oxidant and obtained a copper recovery of 80% after 2 h, highlighting that the flotation tailings should be valorized rather than used for land reclamation. Chen et al. [6] demonstrated that low-grade copper sulfide tailings with pyrite proportions of 31.50%, can be recycled for environmentally friendly disposal through leaching and fractional precipitation methodology, from which a copper recovery of 98.45% could be achieved. Selective extraction of copper from tailings was successively investigated by Turan et al. [21] using a mixture of ammonia salts as leaching reactants and managed to obtain a copper recovery of 91.47% with no iron tenors to the leachate after 6 h of leaching at 30 °C.

This study evaluates the high-pressure oxidative leaching (HPOL) method as an effective way to recover copper and reduce the environmental loading from the mine tailings obtained from Bor Copper Mine, Serbia. The advantages of pressure leaching include fast kinetics, enhanced metal selectivity, and the generation of stable residues [22]. Han et al. [23] have already presented in full detail about high-pressure leaching of the concentrate obtained after the flotation of the mine tailings. This study considers high-pressure leaching directly on the mine tailings as well as environmental evaluation of the leach residue stability. Particular interest was given to the high-temperature oxidation treatment of the mine tailings to convert the reactive and harmful minerals such as pyrite, to more stable forms such as hematite (Fe_2O_3). Chalcopyrite ($CuFeS_2$) has also been listed as one of the sulfide minerals important in AMD formation [5]. Therefore, alongside the conversion of pyrite to hematite, the dissolution of residual chalcopyrite from flotation tailings will also aid in the prevention of AMD. To evaluate the effectiveness of this method in reducing the environmental loading caused by flotation tailings, metal elution tests were performed on the solid residue obtained after HPOL and on the flotation tailings. In addition to Han et al.'s [23] study, this research further investigates the electrowinning process for final copper recovery, in order to demonstrate the feasibility of a complete hydrometallurgical process when utilizing the discarded mine tailings. The electrowinning process was carried out to examine the effect of iron and copper concentration of the stripped solution on current efficiency for copper electrodeposition.

2. Materials and Methods

2.1. Experimental Sample

The mine tailings sample used in this study was obtained from Bor Mine in Serbia. The sample was obtained 14 m below the ground surface of the tailings dam. The average particle size of the mine tailings sample was 27 μm (D_{50}). Table 1 shows the chemical compositions of the two samples utilized in this study; the mine tailings and the concentrate obtained after the flotation of the mine tailings. X-ray diffraction (XRD) analysis (Figure 1) showed that the main mineral compositions of the mine tailings and concentrate of the mine tailings were quartz (SiO_2), pyrite (FeS_2), and kaolinite ($Al_2Si_2O_5(OH_4)$). The main copper mineral in the mine tailings was identified as chalcopyrite ($CuFeS_2$) using scanning electron microscope-energy dispersion spectroscopy (SEM-EDS) (Figure 2). The copper concentrate used in this study was prepared by flotation under determined optimal conditions: pH 10, pulp density 25%, PAX collector dosage 100 g/t-ore, NaHS sulfidizer dosage 1000 g/t-ore, MIBC frother dosage 200 g/t, flotation time 5 min [24]. The obtained copper concentrate and the original mine tailings were both used as feed samples in the subsequent leaching experiments.

Table 1. Chemical compositions of the mine tailings and concentrate.

	Grade (mass%)				
Elements	Cu	Fe	Al	S	SiO$_2$
Mine tailings	0.24	3.51	3.45	4.88	61.7
Concentrate from mine tailings	0.65	33.20	2.63	32.72	23.41

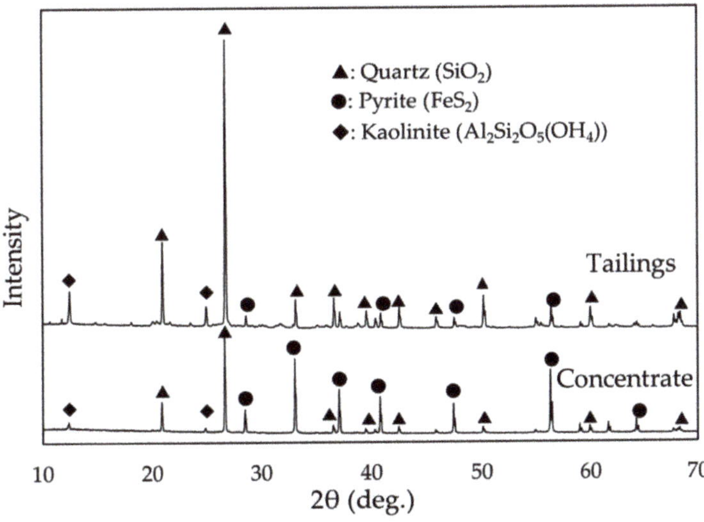

Figure 1. XRD patterns of mine tailings from Bor Copper Mine and concentrate of mine tailings.

Figure 2. SEM image of the mine tailings from Bor Copper Mine.

2.2. Experimental Procedure

2.2.1. High-Pressure Leaching

An autoclave of a 200 mL Teflon beaker was used as a leaching reactor in the high-pressure oxidative leaching experiment (Figure 3) for the mine tailings and concentrate from mine tailings. A sample weight of 10 g was mixed with 0–1.0 M sulfuric acid (H_2SO_4) solution, where 0 M was distilled water, to obtain a pulp density of 100 g/L. The slurry was then placed inside the autoclave and heated. After the temperature reached 180 °C, oxygen gas was injected at 2 MPa into the slurry to provide an oxidative environment. The leaching duration was set to 1 h. After the HPOL experiment, the sample was cooled and filtered to obtain a pregnant leach solution (PLS) and a solid residue, which were taken for chemical and mineralogical analysis using ICP-OES and XRD, respectively.

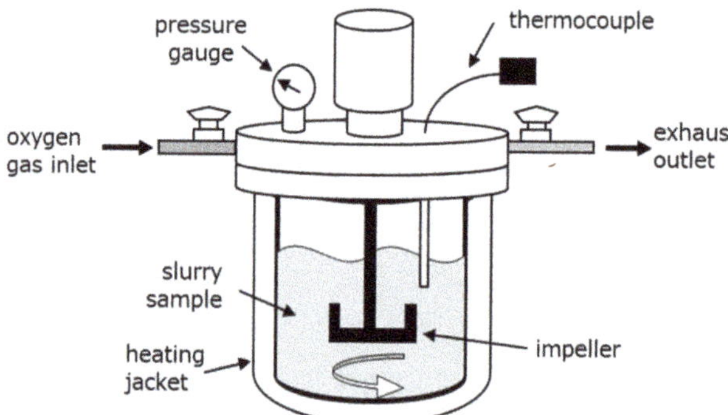

Figure 3. Illustrative setup of an autoclave reactor used for high-pressure oxidative acid leaching.

2.2.2. Elution Test

Elution properties of the mine tailings and the solid residue obtained after HPOL were evaluated. A sample weight of 0.4 g was placed in a sample tube (capacity: 5 mL) together with 4 mL of different pH solutions (pH 2 (adjusted with H_2SO_4), pH 4 (distilled water), and pH 7 (adjusted with $Ca(OH)_2$)). The solutions were shaken for 6 h at 200 rpm, using a shaker (MMS-4020, EYELA). Afterward, the slurry was subjected to solid-liquid separation by centrifugation. Quantitative analysis of each metal in the solution was carried out using ICP-OES (SPS5500, SII nanotechnology). The elution rate was calculated as per Equation (2).

$$\text{Elution rate}(\%) = \frac{C_S * V_S}{C_R * m_R} * 100 \tag{2}$$

where C_S is the concentration of metal in the solution obtained from the elution test (mg/L), and C_R is the concentration of metal in the sample (leach residue/mine tailings), (mg/kg). V_S is the volume of solution (L) and m_R is the dry mass of the sample used (kg).

2.2.3. Solvent Extraction Test

Extraction of copper from the PLS obtained after HPOL of the concentrate from mine tailings was previously carried out by Han et al. [23] using LIX-84I (2-hydroxy-5-nonylace-tophenone oxime) extractant. Kerosene was used to dilute the extractant at a phase ratio of 1. A 10 mL of the PLS and 2 mL of diluted extractant (organic/aqueous phase ratio = 0.2) were placed in a plastic tube and the pH was adjusted to the desired value using a 1.0 M NaOH solution. The solution was mixed using an agitator at 600 rpm and the extraction time was set to 15 min. Afterward, the solution was centrifuged for 5 min at 4000 rpm to ensure a rapid phase separation. To determine the metal extraction efficiency, the raffinate was analyzed using ICP-OES. The Cu-loaded organic obtained from the optimal copper extraction conditions was subjected to stripping experiments where different sulfuric acid concentrations (0.5, 1.0, and 1.5 M) were mixed with the Cu-loaded organic (organic/aqueous phase ratio = 5) at 600 rpm and 15 min contacting time. The stripped solution was analyzed by ICP-OES for the determination of the stripping efficiency and copper concentration [23].

2.2.4. Electrowinning Test

Electrowinning of copper was carried out with a simulated solution prepared in consideration of the copper and ferric ion concentrations of the stripped solution from the solvent extraction stage (44.8 g/L Cu and 1.4 g/L Fe) using distilled water, copper sulfate (II) pentahydrate ($CuSO_4 \cdot 5H_2O$) and iron (III) sulfate n-hydrate ($Fe_2(SO_4)_3 \cdot nH_2O$), where n was analyzed to be 6.33. The electrolyte also contained 170 g/L of free H_2SO_4, and no additives were used. The electrolyte temperature was maintained at 40 °C with a bath heater and the volume of the electrolyte was 0.5 L. The anode and cathode were made from a platinum plate of 1 cm^2 surface area and mounted on epoxy resin with a conducting wire. The anode and cathode were positioned in a fixed mount facing each other, the distance between the electrodes was 3 cm and the mount was placed in the preheated electrolyte cell. A regulated DC power supply (Takasago, Ltd. GP025-5) was used, the operating current density was set at 250 A/m^2, and the electrolysis time was 4 h. To further study the effects of metal concentration in the electrolyte, the copper and iron concentrations were varied between 25–45 g/L and 0–1.5 g/L, respectively. The cathode was weighed before and after the electrolysis test to determine the weight of copper plated. The current efficiency was calculated as per Equation (3). The theoretical mass of the copper deposited was calculated using Equation (4):

$$Current\ Efficiency,\ (C.E) = \frac{mass\ of\ Cu\ deposited\ (actual)}{mass\ of\ Cu\ deposited\ (theoretical)} * 100 \tag{3}$$

$$Cu\ deposited\ (g) = \frac{mm * j * A * t}{nF} \tag{4}$$

where mm is the molar mass of copper, j is the current density, A is the electrode surface area, t is the electrolysis time, n is the number of electrons and F is faradays constant.

3. Results and Discussion

3.1. High-Pressure Leaching

3.1.1. Leaching of Mine Tailing

The effect of sulfuric acid concentration on direct pressure leaching of mine tailings was investigated using distilled water and 0.2–1.0 M H_2SO_4. Other leaching conditions were kept the same as in Han et.al [23] at a fixed pulp density of 100 g/L, total pressure of 2 MPa, leaching temperature of 180 °C, and leaching time of 1 h. As shown in Figure 4, the highest achieved leaching rate of Cu was 98.72% when distilled water was used as a leaching medium, while the lowest achieved leaching rate of Fe was 16.31% at the same leaching conditions. The Cu and Fe concentrations in the leachate were 0.23 g/L and 0.28 g/L, respectively. The high leaching rate of Cu can be attributed to the oxidation of pyrite which is one of the main minerals in the mine tailings. The pyrite was oxidized as per reaction 1 under a total pressure of 2 MPa and leaching temperature of 180 °C, thus generating sulfuric acid [7,9,25]. This was verified by a decline in the pH value from 3.12 before leaching to 0.8 after leaching. The sulfuric acid generated by pyrite oxidation promoted the leaching of chalcopyrite via reaction 5. Han et al. [23] confirmed that the presence of pyrite (FeS_2) in the feed has an efficient effect on copper dissolution. Antonijevic et al. [7] also found that with the increasing concentration of H^+ ions in the leach solution, the dissolution of copper increased. Figure 5 shows the Eh–pH diagram, where the point indicates the pH and Eh (pH: 0.8, Eh: 676 mV) of the solution obtained after leaching. The stability of Cu^{2+} in solution is favored by the leaching conditions of a high temperature and high pressure thus enabling a high leaching rate. XRD measurement of the solid residue (Figure 6) showed the presence of hematite (Fe_2O_3) and the absence of pyrite which was initially in the mine tailings. This indicates that the majority of ferric sulfate is hydrolyzed at a high temperature and total pressure to form hematite as shown by reaction 6. The precipitation of Fe as hematite resulted in a high metal selectivity of dissolved copper over iron when only water was used as a leaching medium.

$$2CuFeS_2 + H_2SO_4 + 8.5O_2 \rightarrow 2CuSO_4 + Fe_2(SO_4)_3 + H_2O \tag{5}$$

$$Fe_2(SO_4)_3 + 3H_2O \rightarrow Fe_2O_3 + 3H_2SO_4 \tag{6}$$

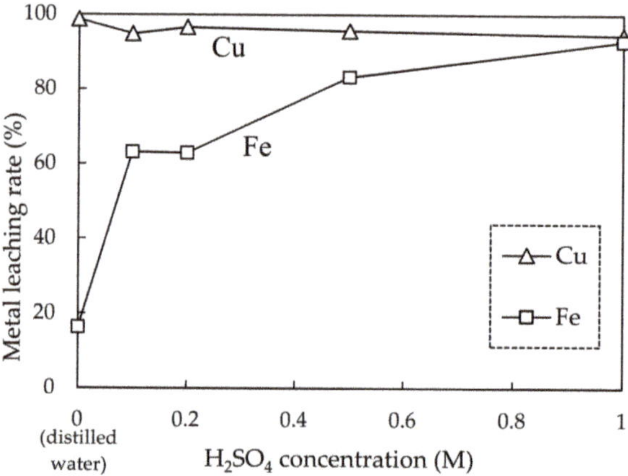

Figure 4. Effect of sulfuric acid concentration on the leaching rate of copper and iron. (Conditions: 100 g/L, 1 h, 700 rpm, 180 °C, 2.0 MPa total pressure).

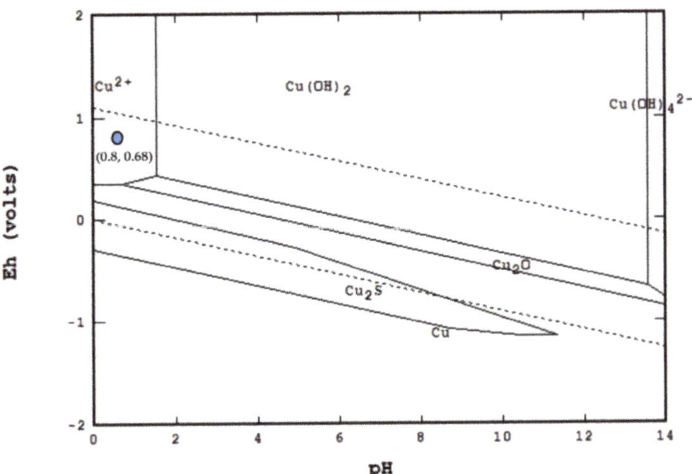

Figure 5. Eh−ph diagram for Cu−S−water system calculated by STABCAL (Condition: 180 °C, 2 MPa).

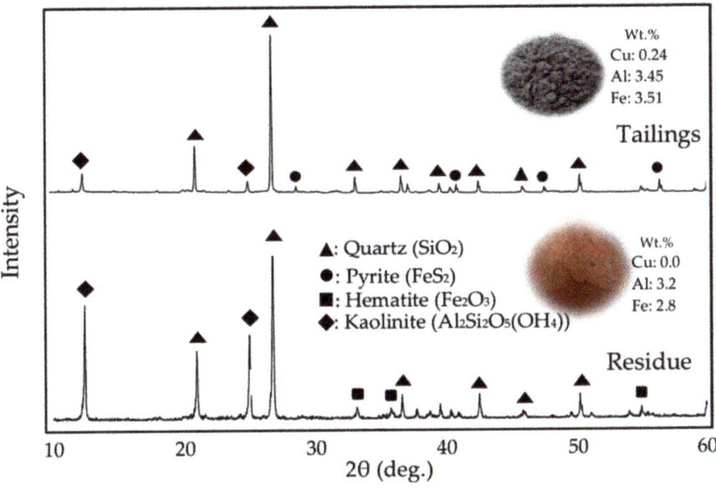

Figure 6. XRD patterns of the residue obtained after leaching of mine tailings with distilled water and the mine tailings.

3.1.2. Elution Test of Leaching Residue Obtained from High-Pressure Leaching

The elution test results are displayed in Figure 7a,b, showing copper and iron concentrations in the solution, respectively. The national effluent standards of copper and iron in Serbia could not be identified, therefore Japan was used for comparison of the results obtained from the elution test. Based on the wastewater discharge standards specified by the Ministry of the Environment in Japan the upper limit for Cu and Fe in the wastewater discharge is 3 mg/L and 10 mg/L, respectively [26]. The concentration of copper in the solution of mine tailings exceeded this criterion at all the investigated pH values (Figure 7a), while the iron concentration in solution exceeded this criterion at a low pH value of 2 (Figure 7b). The elution rate of copper and iron from the mine tailings was calculated to be 4.58% and 0.46% respectively. On the other hand, the solid residue obtained from leaching the mine tailings could meet the regulatory metal criterion with a very low metal concentration below the detection limit (<0.1 ppm) of ICP-EOS. Under various pH conditions, Fe did not elude, mainly because hematite is thermodynamically stable and therefore, less soluble. Furthermore, the elution rate of copper in the solid residue remained undetectable because 98.72% of copper has been recovered in the PLS. It can thus be confirmed that the elution rate of the mine tailings can be reduced by the HPL process, generating a benign solid residue of a reduced environmental loading.

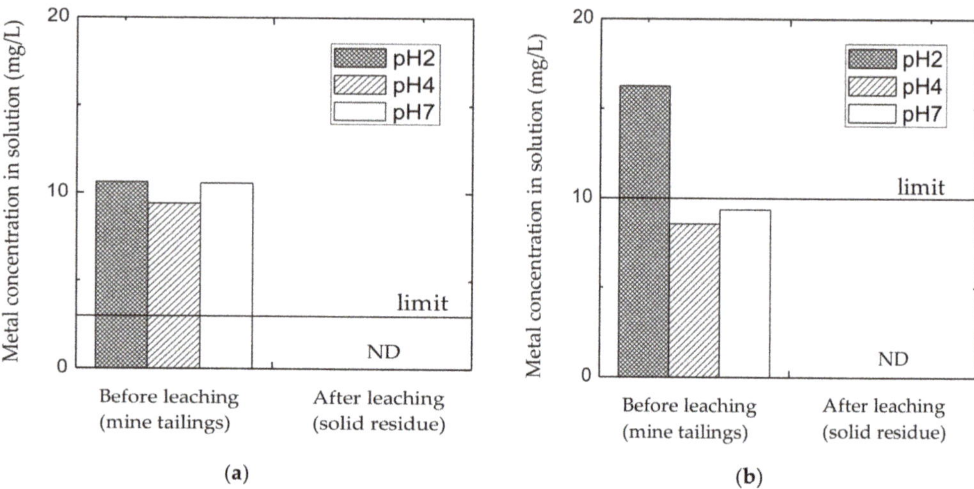

Figure 7. Metal concentration in the solution obtained from the elution test of mine tailings (before leaching) and solid residue (after leaching): (**a**) copper; (**b**) iron. ND = not detected.

3.1.3. Leaching of Concentrate from Mine Tailings

The grade of copper in mine tailings was upgraded from 0.24% to 0.65% through flotation. The concentrate was subjected to high-pressure leaching under the optimal conditions obtained from leaching the mine tailings (leaching medium distilled water, leaching time 1 h and total pressure 2 MPa controlled by O_2 gas, temperature 180 °C) [23]. To concentrate copper by solvent extraction and obtain a solution suitable for electrowinning, a higher copper concentration in the PLS than that obtained from high-pressure leaching of mine tailings (0.23 g/L) is necessary. For this reason, the pulp density was adjusted to 400 g/L. Muravyov and Fomchenko [20], Antonijevic et al. [7], and Muravyov et al., [4] found that the pulp density has no significant effect on the leaching of metals from flotation tailings after obtaining similar values of dissolution degree with various pulp densities. Han et al. [23] achieved a high copper leaching rate of 94.4%, showing no significant change when compared to the leaching rate of Cu from mine tailings (98.72%) at a lower pulp density of 100 g/L. However, increasing the pulp density to 400 g/L increased the leaching rate of iron up to 65.9%. This is attributed to the increase in the added amount of pyrite, which resulted in an increase in free acid concentration in the solution and thus an increase in iron solubility [23]. The obtained PLS consisted of 2.9 g/L of Cu and 102.9 g/L of Fe with a pH of 0.23. For further recovery of copper from the PLS, solvent extraction was carried out.

3.2. Solvent Extraction of the Pregnant Leach Solution from HPOL

Han et al. [23] reported on the results of the extraction of copper from the PLS under the optimal HPOL conditions and extraction conditions stipulated in Section 2.2.3. Figure 8 was drawn based on the results obtained from Han et al. [23], which show that the extraction of copper increased from 10.1 to 93.7% as the pH was increased. However, the iron extraction rate remained below 11.3% yielding a good separation efficiency of 82.4% between copper and iron. The pH of 2.0 was established as an optimal pH for the selective extraction of copper from the PLS.

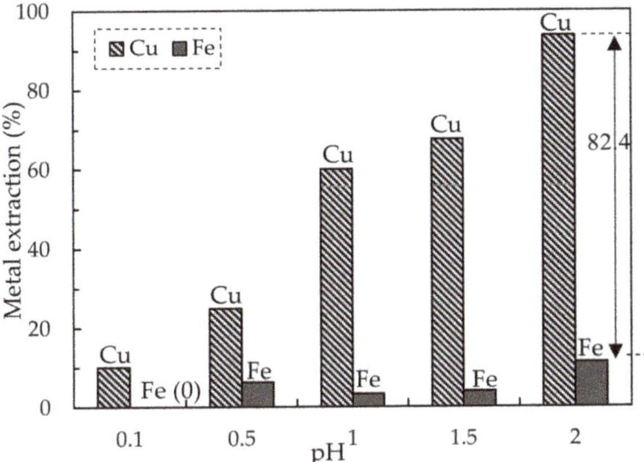

Figure 8. The extraction behavior of metals as a function of pH (Conditions: LIX-84I/Kerosene mixing ratio 1, organic/aqueous phase ratio 0.2, agitating speed 600 rpm, and contact time 15 min).

Han et al. [23] also examined the stripping efficiency of copper from the Cu-loaded organic phase using three different H_2SO_4 concentrations of 0.5, 1.0, and 1.5 M. The highest copper stripping efficiency of 97.4% was obtained at an H_2SO_4 concentration of 1.5 M and was selected as the most suitable H_2SO_4 concentration for stripping copper from the Cu-loaded organic phase. Figure 9 was prepared considering the data from Han et al. [23] of the metal concentrations in stripped solution as a function of H_2SO_4 concentration. As the H_2SO_4 concentration was increased, the copper concentration in the stripped solution increased while the iron concentration decreased. The optimal stripped solution contained approximately 44.8 g/L copper and 1.4 g/L iron [23]. This is comparable to the industrial electrolyte from the solvent extraction stage which typically contains 45 g/L Cu and less than 2.0 g/L of Fe concentration [3] (p. 356).

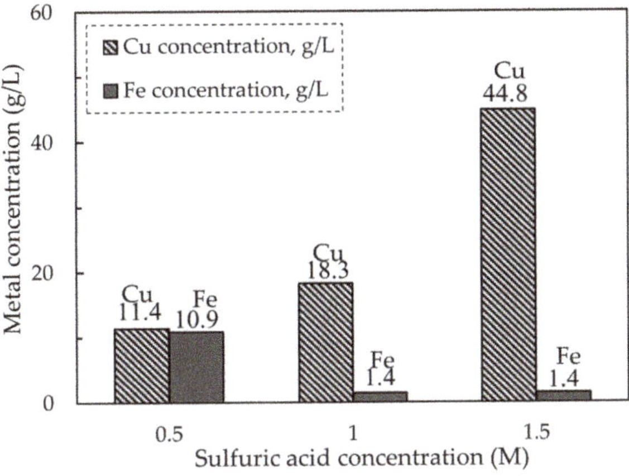

Figure 9. The metal concentration in stripped solution as a function of sulfuric acid concentration. (Conditions: organic/aqueous phase ratio 5, agitation speed 600 rpm, and contact time 15 min).

3.3. Electrowinning of the Simulated Stripped Solution

Current efficiency is an important parameter in electrowinning as it indicates the effectiveness of the electrowinning process. Ferric ion concentration is known to have negative interactions that significantly affect the current efficiency of copper electrodeposition during electrowinning. Therefore, to evaluate the efficiency of carrying out the electrowinning process on the obtained stripped copper solution, it is relevant to examine the current efficiency of the electrowinning process due to the inevitable Fe contamination during leaching. Figure 10 shows the influence of ferric ion concentration on the current efficiency of copper electrodeposition. As expected, a strong inverse relationship is observed, as the ferric ion concentration increases, the current efficiency of copper electrodeposition decreases. This is due to the presence of iron in the copper electrolyte which undergoes reduction from Fe^{3+} to Fe^{2+} at the cathode and re-oxidation to Fe^{3+} from Fe^{2+} at the anode. The cycle proceeds and consumes power that could be used for the deposition of copper. Considering the obtained Fe concentration of 1.4 g/L in the stripped solution, the expected current efficiency is approximately 95%, which is in the upper limit of the industrial range of 85–95% [3] (p. 358). Current efficiency loss was determined to be 2.8% per g/L of ferric ions, which agrees with Khouraibchia and Moats [27] 's empirical model of current efficiency (Equation (7)) and Schlesinger et al., [3] (p. 361)'s findings that current efficiency drops by approximately 2.5% for each addition of 1 g/L of ferric ions.

$$\begin{aligned}
\text{Current efficiency} \quad &(\text{Khouraibchia and Moats 's empirical model}) \\
&= 88.19 - 4.19 * [Fe^{3+}](g/L) + 0.52 * [Cu^{2+}](g/L) \\
&+ 1.81 * 10^{-3} * j(A/m^2) - 6.83 * 10^{-3} * \left[Cu^{2+}\right]^2 (g/L) \\
&+ 0.028 * [Fe^{3+}](g/L) * [Cu^{2+}](g/L) + 4.015 * 10^{-3} \\
&* [Fe^{3+}](g/L) * j\left(A/m^2\right)
\end{aligned} \tag{7}$$

where j is the current density.

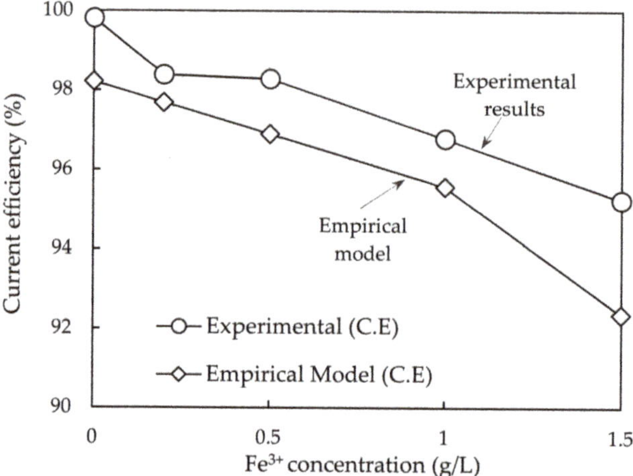

Figure 10. Effect of Fe^{3+} on current efficiency (C.E) of copper deposition. (Conditions: current density 250 A/m^2, temperature 40 °C, electrolysis time 4 h, Cu^{2+} concentration 45 g/L).

Alongside the targeting of a low Fe concentration in the stripped solution, enriching copper concentration was also an important factor during the extraction process. To demonstrate the importance of obtaining a high Cu^{2+} concentration in the stripped solution, the effect of three copper concentrations (25, 35, and 45 g/L) on current efficiency during electrowinning was investigated. The Fe^{3+} concentration in the electrolyte was varied from

0 to 1.5 g/L for each copper concentration investigated. Figure 11 displays the results obtained, showing that current efficiency losses are higher on dilute copper solutions than on concentrated copper solutions. From the gradients of the three plots, current efficiency loss per g/L of Fe^{3+} was determined and is shown in Table 2. A solution of 45 g/L copper concentration had the lowest current efficiency loss. This could be explained by the way that a high Cu^{2+} concentration in the electrolyte constantly provides sufficient copper ions to the cathode surface, thus improving deposition rate as well as the copper current efficiency [28]. Das and Krishna [29] also wrote that increasing the bath Cu^{2+} concentration increases the electrolyte viscosity, which impedes the distribution of Fe^{3+} over the cathode surface. Therefore, in this investigation, obtaining a high copper concentration of 44.8 g/L significantly contributed to achieving a good current efficiency of about 95%.

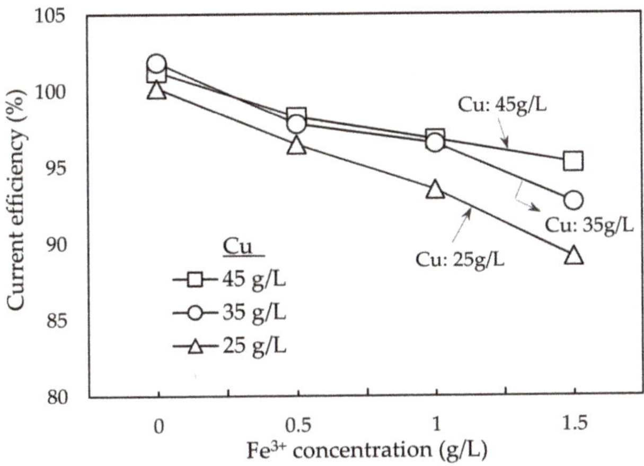

Figure 11. Effect of Fe^{3+} and Cu^{2+} concentration on Cu current efficiency. (Conditions: current density 250 A/m^2, temperature 40 °C, electrolysis time 4 h).

Table 2. The effect of copper concentration on current efficiency loss per g/L of Fe after the electrowinning experiments.

Cu^{2+} Concentration (g/L)	Current Efficiency Losses (Loss per g/L Fe (%))
25	3.62
35	2.88
45	1.95

In hydrometallurgical processes like these, where pyrite is vital to the leaching stage, an Fe contamination in the stripped solution is inevitable. Therefore, obtaining a high copper concentration in the stripped solution is important as it is less susceptible to current efficiency losses during electrowinning. Since the electrowinning experiments were conducted using synthetic solutions, the expected current efficiency on the real solution obtained from solvent extraction may be slightly lower than that obtained when utilizing a synthetic solution. This is due to the complexity of the stripped solution composition because of several other impurities that may cause side reactions. However, current efficiency loss due to ferric ions reduction has been found to have a significant effect on current efficiency compared to other impurities which showed no effect [30]. Therefore, the results obtained from the electrowinning of a simulated solution can be highly comparable to those obtained from electrowinning the real stripped solution under optimized conditions.

4. Conclusions

This study demonstrated that leaching mine tailings could alleviate the possible environmental effects they can pose if left untreated, while simultaneously valorizing them. As an AMD preventative technique, the removal of pyrite from mine tailings through oxidative high-pressure leaching was successively achieved, resulting in the formation of hematite in the solid residue. The obtained residue from leaching the mine tailings, and the original sample of the mine tailings, were subjected to standardized elution tests. The eluded metal concentrations for the solid residue could meet the Ministry of the Environment of Japan's set limits for copper and iron in the wastewater discharge, while the original mine tailings exceeded the set limits, thus confirming the reduced environmental loading of the proposed copper recovery process.

A viable copper recovery process for re-treatment of mine tailings was developed by employing hydrometallurgical methods of high-pressure leaching, solvent extraction, and electrowinning. During high-pressure leaching, the generation of sulfuric acid due to pyrite oxidation promoted copper leaching kinetics when water was used as a leaching medium, thus yielding a high copper leaching rate of 94.4%. Under the optimized solvent extraction conditions, over 93.7% of copper was extracted from the PLS while most of the iron was left in the organic phase. A high copper stripping efficiency of 97.4% was obtained resulting in an enriched solution containing 44.8 g/L Cu and 1.4 g/L Fe. Due to a minimized iron carryover from the stripped solution, electrowinning power consumption by iron was greatly reduced and current efficiency for copper electrodeposition was over 95%.

Author Contributions: Conceptualization, K.H., Y.T., V.T., L.A., Z.S. and A.S.; data curation, L.L.G. and B.A.; formal analysis, L.L.G., K.H., Y.T., D.I., V.T. and A.S.; investigation, L.L.G., K.H., Y.T., V.T. and L.A.; methodology, L.L.G., K.H., Y.T., V.T., L.A., R.J., Z.S. and A.S.; project administration, R.J., Z.S. and A.S.; resources, V.T., L.A., R.J. and Z.S.; supervision, Z.S. and A.S.; validation, D.I.; visualization, K.H.; writing—original draft, L.L.G.; writing—review & editing, L.L.G., K.H., B.A., Y.T., D.I., V.T., L.A., R.J., Z.S. and A.S. All authors have read and agreed to the published version of the manuscript.

Funding: This research was supported by the Leading Program "New Frontier Leader Program for Rare Metals and Resources", the Science and Technology Research Partnership for Sustainable Development (SATREPS), Japan Science and Technology Agency (JST)/Japan International Cooperation Agency (JICA). And the experiments were supported by Akita University Support for Fostering Research Project. The authors greatly acknowledge their financial support.

Institutional Review Board Statement: Not applicable.

Informed Consent Statement: Not applicable.

Data Availability Statement: The data presented in this study are available on request from the corresponding authors.

Conflicts of Interest: The authors declare no conflict of interest. The funders had no role in the design of the study; in the collection, analyses, or interpretation of data; in the writing of the manuscript, or in the decision to publish the results.

References

1. Vriens, B.; Plante, B.; Seigneur, N.; Jamieson, H. Mine Waste Rock: Insights for Sustainable Hydrogeochemical Management. *Minerals* **2020**, *10*, 728. [CrossRef]
2. Tayebi-Khorami, M.; Edraki, M.; Corder, G.; Golev, A. Re-Thinking Mining Waste Through an Integrative Approach Led by Circular Economy Aspirations. *Minerals* **2019**, *9*, 286. [CrossRef]
3. Schlesinger, M.E.; King, M.J.; Sole, K.C.; Davenport, W.G. *Extractive Metallurgy of Copper*, 5th ed.; Elsevier: London, UK, 2011; pp. 68, 356, 358 and 361.
4. Muravyov, M.I.; Bulaev, A.G.; Kondrat'eva, T.F. Complex treatment of mining and metallurgical wastes for recovery of base metals. *Miner. Eng.* **2014**, *64*, 63–66. [CrossRef]
5. Park, I.; Tabelin, C.B.; Jeon, S.; Li, X.; Seno, K.; Ito, M.; Hiroyoshi, N. A review of recent strategies for acid mine drainage prevention and mine tailings recycling. *Chemosphere* **2019**, *219*, 588–606. [CrossRef] [PubMed]

6. Chen, T.; Lei, C.; Yana, B.; Xiao, X. Metal recovery from the copper sulfide tailing with leaching and fractional precipitation technology. *Hydrometallurgy* **2014**, *147*, 178–182. [CrossRef]
7. Antonijevi'c, M.M.; Dimitrijevi'c, M.D.; Stevanovi'c, Z.O.; Serbula, S.M.; Bogdanovic, G.D. Investigation of the possibility of copper recovery from the flotation tailings by acid leaching. *J. Hazard. Mater.* **2008**, *158*, 23–34. [CrossRef]
8. Huang, Z.; Jiang, L.; Wu, P.; Dang, Z.; Zhu, N.; Liu, Z.; Luo, H. Leaching characteristics of heavy metals in tailings and their simultaneous immobilization with triethylenetetramine functioned montmorillonite (TETA-Mt) against simulated acid rain. *Environ. Pollut.* **2020**, *266*, 115236. [CrossRef]
9. Li, X.; Gao, M.; Hiroyoshi, N.; Tabelin, C.B.; Taketsugu, T.; Ito, M. Suppression of pyrite oxidation by ferric-catecholate complexes: An electrochemical study. *Miner. Eng.* **2019**, *138*, 226–237. [CrossRef]
10. Alam, R.; Shang, J.Q. Effect of operating parameters on desulphurization of mine tailings by froth flotation. *J. Environ. Manag.* **2012**, *97*, 122–130. [CrossRef]
11. Guo, Y.; Huang, P.; Zhang, W.; Yuan, X.; Fan, F.; Wang, H.; Liu, J.; Wang, Z. Leaching of heavy metals from Dexing copper mine tailings pond. *Trans. Nonferrous Met. Soc. China* **2013**, *23*, 3068–3075. [CrossRef]
12. Ahmadi, A.; Khezri, M.; Abdollahzadeh, A.A.; Askari, M. Bioleaching of copper, nickel and cobalt from the low grade sulfidic tailing of Golgohar Iron Mine, Iran. *Hydrometallurgy* **2015**, *154*, 1–8. [CrossRef]
13. Figueiredo, J.; Vila, M.C.; Matos, K.; Martins, D.; Futuro, A.; Dinis, M.; Góis, J.; Leite, A.; Fiúz, A. Tailings reprocessing from Cabeço do Pião dam in Central Portugal: A kinetic approach of experimental data. *J. Sustain. Min.* **2018**, *17*, 139–144. [CrossRef]
14. Lü, C.; Wang, Y.; Qian, P.; Liu, Y.; Fu, G.; Ding, J.; Ye, S.; Chen, Y. Separation of chalcopyrite and pyrite from a copper tailing by ammonium humate. *Chin. J. Chem. Eng.* **2018**, *26*, 1814–1821. [CrossRef]
15. Mäkinen, J.; Salo, M.; Khoshkhoo, M.; Sundkvist, J.; Kinnunena, P. Bioleaching of cobalt from sulfide mining tailings; a mini-pilot study. *Hydrometallurgy* **2020**, *196*, 105418. [CrossRef]
16. Falagán, C.; Grail, B.M.; Johnson, D.B. New approaches for extracting and recovering metals from mine tailings. *Miner. Eng.* **2017**, *106*, 71–78. [CrossRef]
17. Kondrat'eva, T.F.; Pivovarova, T.A.; Bulaev, A.G.; Melamud, V.S.; Muravyov, M.I.; Usoltsev, A.V.; Vasil'ev, E.A. Percolation bioleaching of copper and zinc and gold recovery from flotation tailings of the sulfide complex ores of the Ural region, Russia. *Hydrometallurgy* **2012**, *111*, 82–86. [CrossRef]
18. Lutandula, M.S.; Maloba, B. Recovery of cobalt and copper through reprocessing of tailings from flotation of oxidized ores. *J. Environ. Chem. Eng.* **2013**, *1*, 1085–1090. [CrossRef]
19. Urosevic, D.M.; Dimitrijevic, M.D.; Jankovic, Z.D.; Antic, D.V. Recovery of Copper from Copper Slag and Copper Slag Flotation Tailings by Oxidative Leaching. *Physicochem. Probl. Miner. Process.* **2015**, *51*, 73–82. [CrossRef]
20. Muravyov, M.I.; Fomchenko, N.V. Biohydrometallurgical treatment of old flotation tailings of sulfide ores containing non-nonferrous metals and gold. *Miner. Eng.* **2018**, *122*, 267–276. [CrossRef]
21. Turan, M.D.; Orhan, R.; Turan, M.; Nizamoğlu, H. Use of Ammonia Salts in Selective Copper Extraction from Tailings. *Min. Metall. Explor.* **2020**, *37*, 1349–1356. [CrossRef]
22. Stopić, R.S.; Friedrich, B.G. Pressure hydrometallurgy: A new chance to non-polluting processes. *Mil. Tech. Cour.* **2011**, *59*, 29–44. [CrossRef]
23. Han, B.; Altansukh, B.; Haga, K.; Stevanović, Z.; Jonović, R.; Avramović, L.; Urosević, D.; Takasaki, Y.; Masuda, N.; Ishiyama, D.; et al. Development of copper recovery process from flotation tailings by a combined method of high-pressure leaching-solvent extraction. *J. Hazard. Mater.* **2018**, *352*, 192–203. [CrossRef]
24. Han, B.; Altansukh, B.; Haga, K.; Stevanovi´c, Z.; Radojka, J.; Markovic, R.; Avramovic, L.; Obradovic, L.; Takasaki, Y.; Masuda, N.; et al. Copper upgrading and recovery process from mine tailing of Bor region, Serbia using flotation. *Soc. Mater. Eng. Resour. Jpn.* **2014**, *20*, 225–229. [CrossRef]
25. McDonald, R.G.; Muir, D.M. Pressure oxidation leaching of chalcopyrite. Part, I. Comparison of high and low temperature reaction kinetics and products. *Hydrometallurgy* **2007**, *86*, 191–205. [CrossRef]
26. Ministry of the Environment, Government of Japan. Uniform National Effluent Standards (Last Update: 21 October 2015). Available online: https://www.env.go.jp/en/water/wq/nes.html (accessed on 30 March 2021).
27. Khouraibchia, Y.; Moats, M. Evaluation of copper electrowinning parameters on current efficiency and energy consumption using surface response methodology. In Proceedings of the 217th ECS Meeting, Vancouver, BC, Canada, 25–30 April 2010.
28. Owais, A. Effect of electrolyte characteristics on electrowinning of copper powder. *J. Appl. Electrochem.* **2009**, *39*, 1587–1595. [CrossRef]
29. Das, S.C.; Gopala Krishna, P. Effect of Fe (III) during copper electrowinning at higher current density. *Int. J. Miner. Process.* **1996**, *46*, 91–105. [CrossRef]
30. Moats, M. How to evaluate current efficiency in copper electrowinning. In *Separation Technologies for Minerals, Coal, and Earth Resources*; Young, C.A., Luttrell, G.H., Eds.; Society for Mining, Metallurgy, and Exploration, Inc. (SME): Englewood, CO, USA, 2012; pp. 333–339.

Article

The Effects of Coexisting Copper, Iron, Cobalt, Nickel, and Zinc Ions on Gold Recovery by Enhanced Cementation via Galvanic Interactions between Zero-Valent Aluminum and Activated Carbon in Ammonium Thiosulfate Systems

Sanghee Jeon [1,*], Sharrydon Bright [2], Ilhwan Park [1], Carlito Baltazar Tabelin [3], Mayumi Ito [1] and Naoki Hiroyoshi [1,*]

[1] Division of Sustainable Resource Engineering, Faculty of Engineering, Hokkaido University, Sapporo 060-8628, Japan; i-park@eng.hokudai.ac.jp (I.P.); itomayu@eng.hokudai.ac.jp (M.I.)

[2] Department of Mining, Chemical and Metallurgical Engineering, Faculty of Engineering, University of Zimbabwe, Harare 00263, Zimbabwe; sharrydonbright@frontier.hokudai.ac.jp

[3] School of Minerals and Energy Resource Engineering, The University of New South Wales, Sydney, NSW 2052, Australia; c.tabelin@unsw.edu.au

* Correspondence: shjun1121@eng.hokudai.ac.jp (S.J.); hiroyoshi@eng.hokudai.ac.jp (N.H.); Tel.: +81-11-706-6918 (S.J.)

Citation: Jeon, S.; Bright, S.; Park, I.; Tabelin, C.B.; Ito, M.; Hiroyoshi, N. The Effects of Coexisting Copper, Iron, Cobalt, Nickel, and Zinc Ions on Gold Recovery by Enhanced Cementation via Galvanic Interactions between Zero-Valent Aluminum and Activated Carbon in Ammonium Thiosulfate Systems. Metals 2021, 11, 1352. https://doi.org/10.3390/met11091352

Academic Editor: Felix A. Lopez

Received: 30 July 2021
Accepted: 25 August 2021
Published: 27 August 2021

Abstract: The use of galvanic interactions between zero-valent aluminum (ZVAl) and activated carbon (AC) to recover gold (Au) ions is a promising technique to overcome the challenges due to the poor recovery in ammonium thiosulfate systems, but the applicability to practical Au ore processing remains elusive so far. The present study describes (1) the recovery of Au ions from low Au concentrations, which are typical concentrations used in Au ore processing; and (2) an investigation into the effects of various coexisting base metal ions that can be present in pregnant ore-leached solutions. The results showed that high Au recovery (i.e., over 85%) was obtained even at low Au concentrations under the following conditions: 1:1 of 0.15 g of ZVAl and AC with 10 mL of ammonium thiosulfate solution containing 5 mg/L of Au ions at 25 °C for 1 h in an anoxic atmosphere. Selected coexisting metal ions (i.e., copper, iron, cobalt, nickel, and zinc) were studied to establish their effects on Au recovery, and the results showed that the Au recovery was enhanced (about 90%) when copper ions coexist in the solution with minimal effects from other competing base metal ions.

Keywords: ammonium thiosulfate; gold; cementation; galvanic interaction; zero-valent aluminum; activated carbon

1. Introduction

Changing international regulations, consumer perceptions, and investor expectations in recent years have pushed for more sustainable and eco-friendly mineral processing and metal extraction technologies [1,2]. In gold (Au) hydrometallurgy, cyanide-based technologies such as carbon-in-pulp (CIP) and carbon-in-leach (CIL) remain widely used in both medium- and large-scale Au mining operations [3]. Unfortunately, cyanide is a very toxic compound that poses serious environmental and health hazards when improperly handled and disposed of, so its use is strictly controlled and prohibited in many countries [1,4].

Among the many alternative methods for Au extraction, copper (Cu)-catalyzed ammonium thiosulfate leaching is one of the most promising because it uses lixiviants that are non-toxic and less corrosive (Equation (1)) [4,5].

$$Au + 5S_2O_3^{2-} + Cu(NH_3)_4^{2+} \rightarrow Au(S_2O_3)_2^{3-} + 4NH_3 + Cu(S_2O_3)_3^{5-} \tag{1}$$

Furthermore, this approach is effective for the treatment of secondary resources such as e-wastes that contain various types of materials such as plastics, resins, and ferrous, base, and precious metal alloys. Ha et al. [6], for example, reported that over 98% of Au was leached from waste mobile phones. Similarly, Jeon et al. [7] successfully leached 99% of Au from the printed circuit boards (PCBs) of waste mobile phones under the following conditions: 1 M of thiosulfate, 1 M of ammonia/ammonium, and 10 mM of Cu ions for 24 h at 25 °C. Another potential application of Cu-catalyzed ammonium thiosulfate leaching is in the treatment of carbonaceous-type and pyritic Au ores that are unsuitable for cyanide-based methods with the absence of preg-robbing. As a result, remarkable leaching studies for Au ores also have been reported as follows: Molleman and Dreisinger [8] extracted about 84% of Au from pyritic Au concentrate after 24 h using this method, while Ficeriova et al. [9] dissolved 99% of Au within 45 min with complex sulfide concentrates. Despite these promising results, applications of Cu-catalyzed ammonium thiosulfate leaching in industrial-scale plants remain limited to date because an acceptable method for Au ion recovery from pregnant leach liquors/solutions remains elusive [4,7].

Conventional CIP and CIL technologies employ activated carbon (AC) to recover Au ions from pregnant leach solutions, an approach that is simple and highly efficient [1,10]. Although activated carbon is very efficient when used with cyanide, it is ineffective when thiosulfate is employed because of the low adsorption affinity of the larger, more negative Au thiosulfate complex adsorption to the AC [1]. According to Navarro et al. [11], Au recovery by AC adsorption from an ammonium thiosulfate medium was only about 50% after 8 h. Cementation (reductive precipitation), an electrochemical process whereby Au ions are reduced to metallic Au by reductants, is also a well-established recovery technique for cyanide-based technologies. Metal reductants or cementation agents such as zero-valent base metals (e.g., copper (Cu), zinc (Zn), aluminum (Al), and iron (Fe)) are often used. In ammonium thiosulfate, however, cementation of Au ions is difficult because of various unwanted side reactions [7,12–15]: (1) reduction of Cu ions employed as a catalyst; (2) dissolution of the cementation agents, which leads to high reagent consumption; and (3) formation of oxide/sulfide layers on cementation agents that inhibit Au recovery. Furthermore, abundant sulfur and Cu ions present in the solution restrict the application of solvent extraction and electrowinning for Au ion recovery [1,15].

In our previous study, a novel recovery technique that uses synergistic interactions between zero-valent aluminum (ZVAl) and AC for enhanced Au recovery in the ammonium thiosulfate system was developed [15]. The results showed that ZVAl or AC alone could not recover Au ions, consistent with the results of many previous studies. When mixed, however, over 99% of Au ions could be recovered through the following mechanisms: (i) ZVAl acts as an electron donor while AC as an electron mediator to an Au thiosulfate complex; and (ii) making a galvanic cell, which finally leads to enhanced Au recovery [15]. Although the previous study established a high recovery of Au ions from ammonium thiosulfate solutions, this was obtained in a model solution, which contains only a high concentration of Au ions. This means that the applicability of this technique to real Au ore processing and/or e-waste recycling with a low Au ion concentration and coexisting metal ions remains untested.

The present study aims to assess the applicability of this simple and highly efficient novel recovery technique (ZVAl-AC recovery technique) to real Au ore processing. The objectives of this paper are specifically as follows: (a) to recover Au from the solution with less than 10 mg/L Au, and (b) to investigate the effects of various coexisting metal ions on Au recovery. The previous study on Au ion recovery by the ZVAl-AC technique obtained high Au recovery (i.e., 99%) from the solutions but contained a 100 mg/L concentration of Au ions, which is much higher than the typical Au concentration in ores. There are surely mines containing high Au concentrations such as 150 g/t in Australia [16], 94 g/t in Korea [17], and 60 g/t in China [18], but those mines with high Au contents have been actively explored; hence, currently operating/investigating mines mainly deal with refractory or complex ores with relatively low Au concentrations such as 6 g/t in Laos [19],

6.2 g/t in China [20], 6.2 g/t in Iran [21], or 11.2 g/t in Ghana [22]. Furthermore, Au ore contains minerals such as pyrite, arsenopyrite, chalcopyrite, and/or malachite in which various elements (e.g., Cu, Fe, Co, Ni, and Zn) are incorporated [23–28]. Once these elements are dissolved in the solution, they can affect Au recovery by competing and/or co-depositing with Au during the recovery process. Thus, identification of Au recovery mechanisms in solutions with low Au concentrations and coexisting metal ions will be essential in the industrial-scale application of ammonium thiosulfate leaching.

The first objective was achieved by batch-type experiments in ammonium thiosulfate solution containing only Au ions (i.e., Au-thiosulfate solution) at low concentrations (i.e., 5 and 10 mg/L), while for the second objective, different concentrations of coexisting metal ions (i.e., Cu, Fe, Co, Ni, or Zn ions) were added to the Au-thiosulfate solutions with and without Cu ions. The residues were also observed using a scanning electron microscope with energy dispersive X-ray spectroscopy. This study provides a bridge for the application of this simple, highly efficient recovery technique, which has remained in model testing for Au ore processing, and will be helpful for researchers interested in the recovery of Au ions as well as in the behavior of base metal ions in thiosulfate systems.

2. Materials and Methods

2.1. Recovery of Au Ions from Solutions with a Low Au Concentration

The stock ammonium thiosulfate solution containing Au ions (i.e., Au-thiosulfate solution) was prepared by dissolving 5 or 10 mg/L of Au powder (99.999%, Wako Pure Chemical Industries, Ltd., Osaka, Japan) in an ammonium thiosulfate solution containing 1 M of $Na_2S_2O_3$, 0.5 M of NH_3, 0.25 M of $(NH_4)_2SO_4$, and 10 mM of $CuSO_4$ (pH between 9.5 and 10) using a 300 mL Erlenmeyer flask shaken in a thermostat water bath shaker at 25 °C for 24 h with constant shaking at 120 min^{-1}. The concentration of Au ions in the stock solution was measured by inductively coupled plasma atomic emission spectroscopy (ICP-AES, ICPE-9820, Shimadzu Corporation, Tokyo, Japan) (margin of error = ±2%). Subsequently, 0.15 g of ZVAl (99.99%, Wako Pure Chemical Industries, Ltd., Osaka, Japan) and/or 0.15 g of AC (99.99%, Wako Pure Chemical Industries, Ltd., Osaka, Japan) was mixed with 10 mL of Au-thiosulfate solution in 50 mL Erlenmeyer flasks at 25 °C (shaking at 120 min^{-1}) under nitrogen purging conditions to remove the dissolved oxygen in the solution. After 1 h, the filtrate and residue were separated by filtration using 0.2 μm syringe-driven membrane filters (LMS Co., Ltd., Tokyo, Japan). The residues were washed thoroughly with deionized water (18 MΩ·cm, Mill-Q® Integral Water Purification System, Merck Millipore, Billerica, MA, USA), dried in a vacuum oven at 40 °C, and analyzed by a scanning electron microscope with energy-dispersive X-ray spectroscopy (SEM-EDX, Superscan SSX-550, Shimadzu Corporation, Tokyo, Japan). Meanwhile, the concentrations of Au ions remaining in the filtrates were analyzed by ICP-AES. For validity, accuracy, and replicability of results, experiments were conducted in triplicates.

2.2. Recovery of Au Ions from Solutions Containing Coexisting Metal Ions

In order to investigate the effects of coexisting metal ions on Au recovery using the ZVAl-AC technique, Au-thiosulfate solutions together with base metal ions (Cu, Fe, Co, Ni, and Zn ions) of concentration varying from 0 to 50 mM with and without Cu ions were prepared. The reagents used as sources of competing ions were analytical-grade powders of $CuSO_4 \cdot 5H_2O$, $NiSO_4 \cdot 6H_2O$, $FeSO_4 \cdot 7H_2O$, $ZnSO_4 \cdot 7H_2O$, and $CoSO_4 \cdot 7H_2O$ (Wako Pure Chemical Industries Ltd., Osaka, Japan). Subsequently, 10 mL of Au-thiosulfate solution containing each base metal ion was purged with ultra-pure N_2 to remove any dissolved oxygen, and then mixed with 0.15 g of ZVAl and AC at 25 °C for 1 h, shaking at 120 min^{-1}. After the predetermined mixing time, the suspension was filtered and the residues were washed thoroughly with deionized water, dried in a vacuum oven at 40 °C, and analyzed by SEM-EDX. Meanwhile, the filtrates were analyzed by ICP-AES. For validity, accuracy, and replicability of results, experiments were conducted in triplicates.

3. Results

3.1. Recovery of Au Ions from Solutions Containing a Low Au Concentration

Figure 1 shows the recovery of Au ions from the ammonium thiosulfate solution containing Au concentrations of 5, 10, and 100 mg/L by ZVAl and AC. Au recovery was calculated according to the following equation:

$$\text{Au recovery (\%)} = \frac{[Au](i) - [Au](f)}{[Au](i)} \times 100\% \tag{2}$$

where $[Au](i)$ and $[Au](f)$ are the initial and final Au concentrations, respectively.

(a)

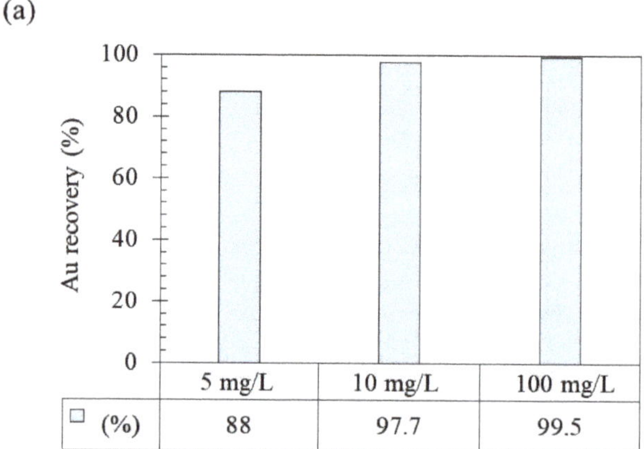

	5 mg/L	10 mg/L	100 mg/L
□ (%)	88	97.7	99.5

(b)

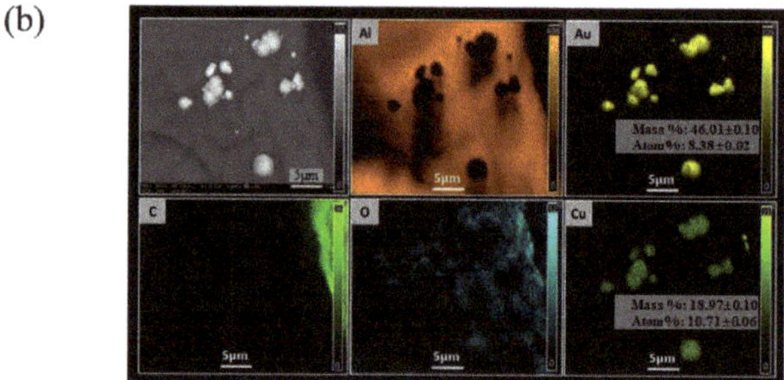

Figure 1. (**a**) Recovery of Au ions from ammonium thiosulfate solutions with varying initial concentrations from 5 to 100 mg/L, and (**b**) SEM photomicrographs with corresponding elemental maps of the recovery residue at 10 mL/L of Au initial concentration.

The results showed that over 99% of Au was recovered when the initial concentration was 100 mg/L, while about 90% of Au was recovered when the initial concentration was 5 mg/L. The results indicate that recovery slightly decreases as the initial concentration decreases, similar to the results reported by Wang et al. [29] and Nguyen et al. [30] for the cementation of Au ions, but still showed a high recovery of 90%. Figure 1b shows the SEM photomicrographs with corresponding elemental maps of the residue at 10 mL/L of Au initial concentration, and the results showed that Au and Cu were detected on the surface of ZVAl.

3.2. Recovery of Au Ions from Solutions Containing Coexisting Metal Ions

3.2.1. Recovery of Au Ions with Varying Cu Concentrations in the Solution

This section focuses on the effects of Cu ions on Au recovery for the following reasons:

- Cu ions are an essential catalyst in ammonium thiosulfate systems, increasing Au dissolution in ammonium thiosulfate systems 20- to 25-fold (Equation (1)) [1,4];
- Cu ions could be introduced via the dissolution of Cu minerals such as chalcopyrite and malachite [23–28] found in Au ores.

Figure 2 presents how Cu ions affect Au recovery when using ZVAl and AC. The results show that the initial increase in Cu ion concentration by 10 mM enhanced Au recovery. This could be attributed to the formation of galvanic cells as a consequence of Cu cementation. Galvanic interactions enhance electron transfer from ZVAl to the Au thiosulfate complex, increasing Au recovery [15]. In the second region from 10 to 40 mM Cu, Au recovery was high and consistent at about 90%. In this region, all the Au that can be recovered was already cemented on the ZVAl; hence, increasing Cu ions did not affect Au recovery. In the final region above 40 mM Cu, excess Cu ions began to reduce Au recovery while Cu precipitation still increased. As shown in Figure 3, the decrease in Au recovery could be attributed to the competition between the reduction of Cu species (Equations (3) and (4)) and Au species (Equation (5)) [4,7,12].

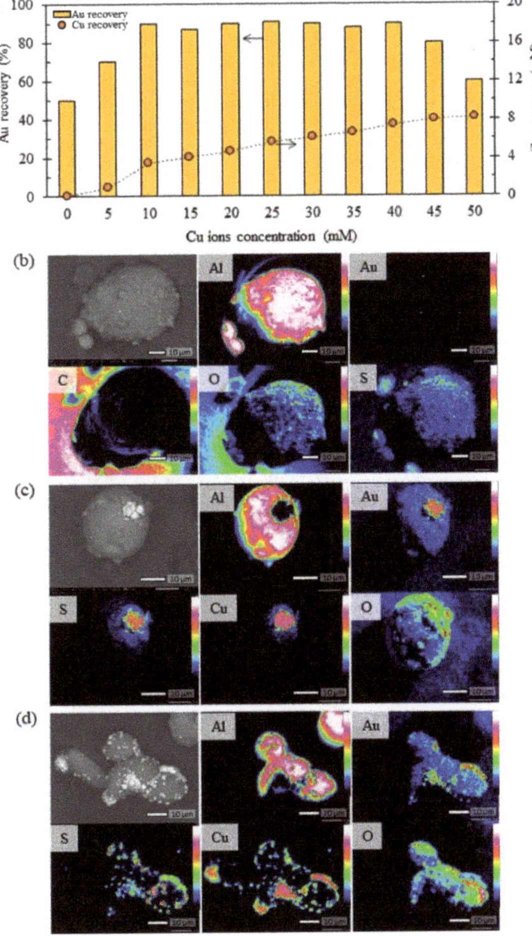

Figure 2. (a) Recovery of gold ions from ammonium thiosulfate solution with varying additions of Cu ions with concentrations from 0 to 50 mM, and SEM photomicrographs with corresponding elemental maps of the residues at (b) 0 mM, (c) 10 mM, and (d) 50 mM of Cu ion additions.

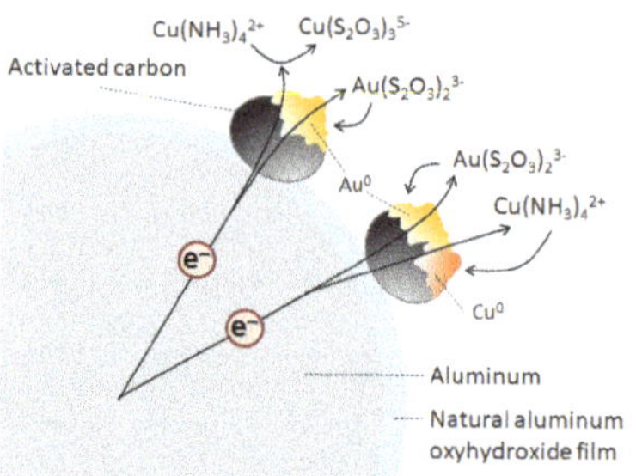

Figure 3. Schematic diagram of the competitive electron transfer from ZVAl to the Cu complex and the Au-thiosulfate complex.

SEM-EDX results showed that in the absence of Cu ions (residue at 0 mM Cu), Au was recovered as small point-like depositions on the surface of ZVAl (Figure 2b), while the area of Au deposition increased when Cu concentration increased by 10 and 50 mM (Figure 2b,c), and showed that deposited Au occurs together with deposited Cu.

$$Cu(NH_3)_4{}^{2+} + 3S_2O_3{}^{2-} + e^- \rightarrow Cu(S_2O_3)_3{}^{5-} + 4NH_3 \tag{3}$$

$$Cu(NH_3)_4{}^{2+} + 2e^- \rightarrow Cu + 4NH_3 \tag{4}$$

$$Au(S_2O_3)_2{}^{3-} + e^- \rightarrow Au^0 + 2S_2O_3{}^{2-} \tag{5}$$

3.2.2. Recovery of Au Ions with Varying Fe, Co, Ni, and Zn Concentrations in the Solution

In this part of the study, various metal ions including Fe, Co, Ni, and Zn ions that could coexist in pregnant Au ore solutions were selected and their effects on Au recovery were elucidated.

The effects of Fe ions on Au recovery are shown in Figure 4. Figure 4a,b shows the recovery of Au ions with varying addition concentrations of Fe ions in the absence and presence of Cu ions in the solution, respectively. In both cases, the results showed that the effects of Fe ions on Au recovery were negligible. This could be explained by the very low solubility of Fe ions under basic conditions [31]; hence, Fe precipitates are readily formed in the solution, which neither favors nor hinders Au recovery. The precipitate in the solution was analyzed by SEM-EDX and showed that Fe precipitates were formed in the solution as shown in Figure 4c. When 10 mM of Cu ions were added to the solution, Au recovery was enhanced and consistently showed about 85–90% Au recovery (Figure 4b), which indicates that Au can be successfully recovered regardless of the presence of Fe ions.

Figure 5a,b shows the results of Au recovery with varying Co ion additions without and with Cu ions, respectively. In the absence of Cu ions (Figure 5a), Au recovery increased to 60% with the addition of 1 mM of Co ions, then decreased as the addition of Co ions increased due to the competitive reduction of Co ions and Au ions. In the presence of Cu ions in the solution, however, Au recovery was about 90%, indicating that the negative effects of Co ions on Au recovery were hindered in the presence of Cu ions and high Au recovery could be obtained (Figure 5b). Figure 5c presents the residue analysis by SEM-EDX. The results show that Co, Au, and Cu were recovered on the surface of ZVAl and deposition areas of Au were very close to that of Cu.

Figure 6a,b shows the effects of Ni ions on Au recovery with varying concentrations without and with Cu ions, respectively. The results showed that in the absence of Cu ions, Au recovery increased to 83% with the addition of 10 mM of Ni ions, then recovery decreased as the addition of Ni ions increased by the competitive reduction. In the presence of Cu ions, however, Au recovery was constantly high at all ranges of Ni ion additions [32]. Figure 6c presents the SEM-EDX analysis of the residue, showing that Ni, Cu, and Au were co-cemented on ZVAl particles.

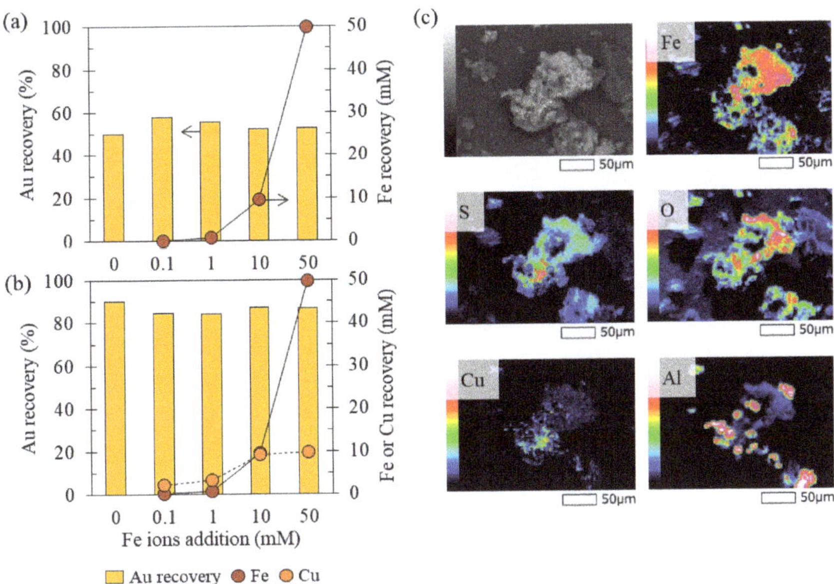

Figure 4. Recovery of gold ions from an ammonium thiosulfate solution with varying addition of Fe ion concentrations from 0 to 50 mM (**a**) without and (**b**) with 10 mM of Cu ions in the solution, and (**c**) SEM photomicrographs with corresponding elemental maps of the precipitates in the solution.

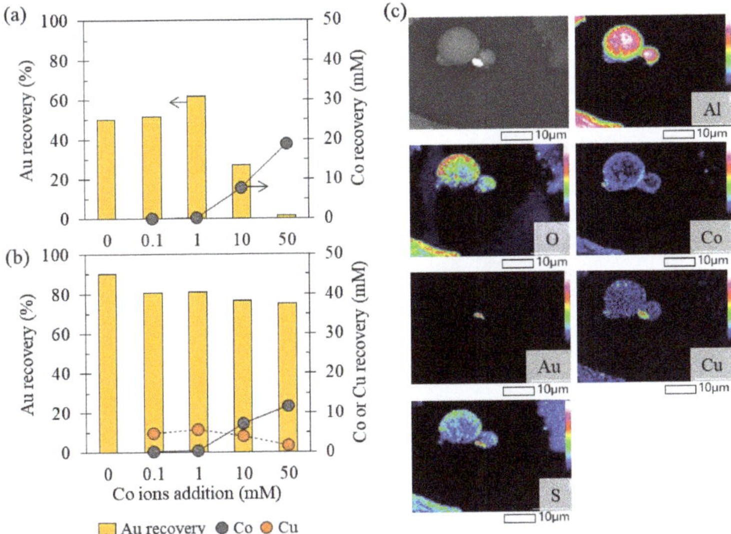

Figure 5. Recovery of gold ions from ammonium thiosulfate solutions with varying additions of Co ion concentrations from 0 to 50 mM (**a**) without and (**b**) with 10 mM of Cu ions in the solution, and (**c**) SEM photomicrographs with corresponding elemental maps of residue.

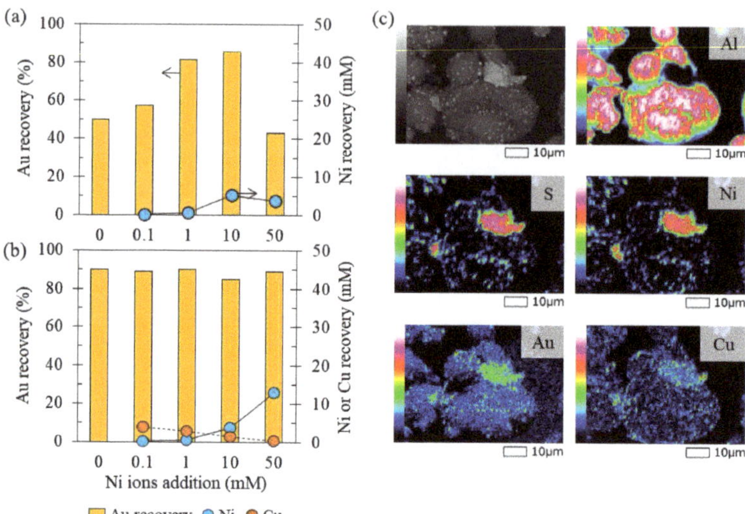

Figure 6. Recovery of gold ions from ammonium thiosulfate solution with varying additions of Ni ion concentrations from 0 to 50 mM (**a**) without and (**b**) with 10 mM of Cu ions in the solution, and (**c**) SEM photomicrographs with corresponding elemental maps of the residue.

Figure 7a,b shows the recovery of Au ions with varying additions of Zn ions in the solution from 0.1 to 50 mM without and with Cu ions in the solution, respectively. Without Cu ions, Au recovery increased with the increase in Zn ion addition, while Au recovery remained at an almost constant value of 90% with Cu ions, regardless of the addition of Zn. The SEM photomicrograph with corresponding elemental maps shows that both Zn and Cu were recovered on the surface of ZVAl with Au.

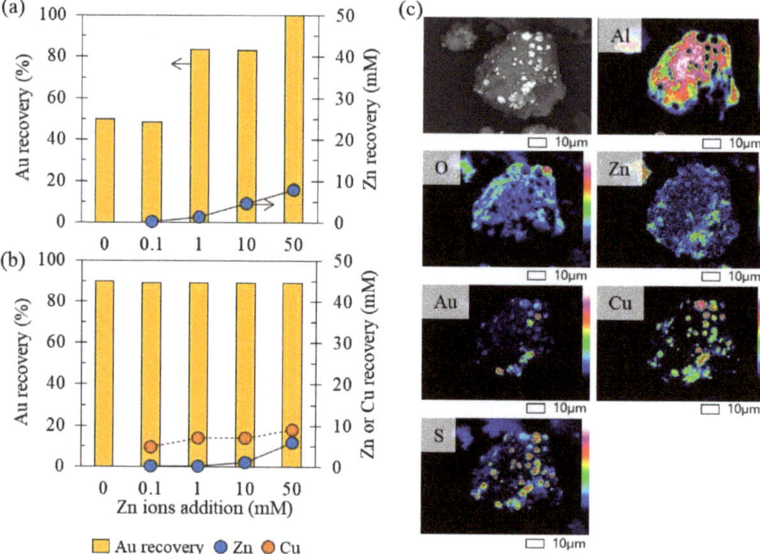

Figure 7. Recovery of gold ions from an ammonium thiosulfate solution with varying additions of Zn ion concentrations from 0 to 50 mM (**a**) without and (**b**) with 10 mM of Cu ions in the solution, and (**c**) SEM photomicrographs with corresponding elemental maps of residue.

The observed results can be summarized by the following four effects with Figure 8:

- Effect 1: Fe ions precipitated from solution phase, and they did not affect Au recovery.
- Effect 2: Zn ions and low concentrations of Co and Ni ions enhanced Au recovery.
- Effect 3: Higher concentrations of Co and Ni ions suppressed Au recovery.
- Effect 4: In the presence of Cu ions, the effects of other coexisting metal ions were hindered, i.e., Au recovery was almost constant (the orange section illustrated in Figure 8), regardless of the presence of other coexisting metal ions.

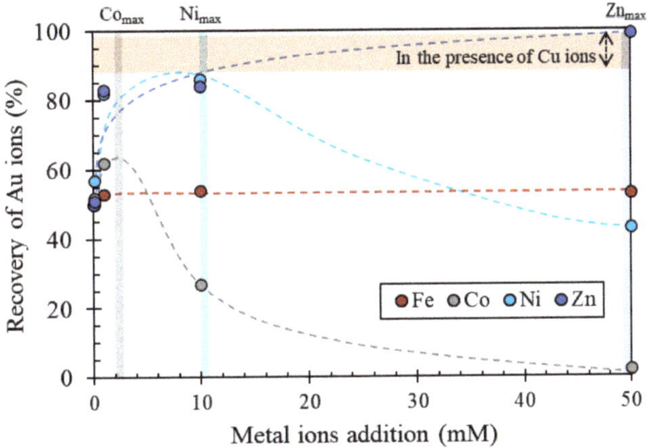

Figure 8. The competitive cementation tendency between Au ions and other base metal ions reacted on the surface of AC attached to ZVAl in the absence of Cu ions.

For Effect 1, Fe ions had low solubility and precipitated in the current system and had no effect on Au recovery. Effect 2 (enhanced Au recovery with some of the metal ions) may have occurred due to the formation of new reaction sites for Au deposition: the coexisting metal ions were reduced on the AC (electron mediator) attached to ZVAl, and the deposit elemental metals acted as secondary electron mediators or new cathode sites for Au deposition. Effect 3 (suppressive effect of the coexisting metal ions on Au recovery) can be interpreted by assuming the competition of cementation reactions (reduction of metal ions) between Au and coexisting metal ions: a limited amount of electron flow from the donor (ZVAl) was shared with both the Au and coexisting metal ions, causing a decrease in the amount of Au deposition. At a certain concentration of metal ions, the enhanced effect (Effect 2) shifted to suppressive effect (Effect 3) and Au recovery reached its maximum at that point. The concentrations of coexisting metal ions promoting maximum Au recovery are shown as Me$_{max}$ in Figure 8. The Me$_{max}$ concentration was lowest for Co, followed by Ni, then highest for Zn. This order may correspond to the order of the standard redox potential of these metal ions, as shown in Table 1: the standard redox potential for $Co(NH_3)_6^{3+}/Co$ is higher than that of the others, suggesting that reductive deposition of Co occurs at lower concentrations. This could be the reason why the enhanced Au cementation occurred at low concentrations of Co. For the Ni and Zn, the standard redox potentials are lower than Co, and higher concentrations are needed to form the new cathode site for Au deposition; hence, higher concentrations of Ni and Zn were needed for the enhanced Au recovery.

Table 1. Standard redox potentials of Au, Cu, Co, Ni, and Zn in ammonium thiosulfate systems [33–37].

Metals	E^0/V
$Au(S_2O_3)_2^{3-}/Au$	0.27
$Cu(NH_3)_4^{2+}/Cu\,(S_2O_3)_2^{3-}$	0.22
$Co(NH_3)_6^{3+}/Co(NH_3)_x^{2+}$ (x is mainly 5 under the current conditions)	0.21
$Co(NH_3)_6^{3+}/Co$	0.1
$Cu(NH_3)_4^{2+}/Cu$	−0.05
$Cu(NH_3)_4^{2+}/Cu_2S$	−0.2
$Ni(NH_3)_6^{2+}/Ni$	−0.49
$Zn(NH_3)_4^{2+}/Zn$	−1.04

From an engineering viewpoint, Effect 4 is probably the most noteworthy result here. In the presence of Cu ions, the effects of other coexisting metal ions were hindered, and high Au recovery was achieved (i.e., 85–95%). As show in Table 1, with the exception of Co ions, standard redox potentials of Cu ions are higher than other metal ions [37,38]. When a high concentrations of Cu ions coexists, Cu ions would preferentially be reduced and deposited on AC/ZVAl surfaces [39]. Cu is an electro-conductor, and its conductivity is much higher than that of other metals (Table 2), and it acts as a high-efficiency secondary electron mediator or cathode site, enhancing electron transfer from ZVAl to Au ions (Figure 9). For that, the effects of other coexisting metal ions (especially the suppressive effect of the ions) on Au recovery would be minimized in the presence of Cu ions, and this is schematically illustrated in Figure 9.

Table 2. Conductivity of Cu, Zn, Co, and Ni [40,41].

Metals	Conductivity (S/m $\times\ 10^7$ at 25 °C)
Cu	5.98
Zn	1.7
Co	1.6
Ni	1.4

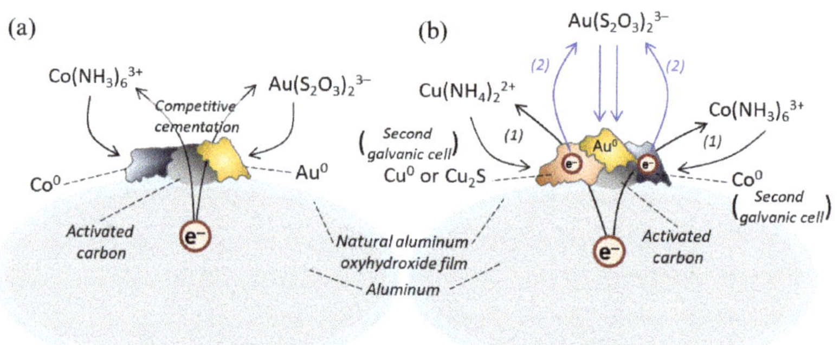

Figure 9. Schematic diagram of the competitive electron transfer from ZVAl to Cu-ammine, Co-ammine, and Au-thiosulfate complex (**a**) without and (**b**) with Cu ions in the system.

3.2.3. Recovery of Au Ions from Solutions Containing Various Coexisting Metal Ions

Figure 10 shows the results of the experiment using the solution containing 10 mM of all base metal ions together with Au ions, like a model solution of pregnant leached solution. The results showed that over 90% of Au was recovered together with Co, Cu, Zn, and Ni, except for Fe ions, and SEM-EDX analysis of the solid product showed that Au and other base metals were deposited on the same sites. This result confirms that high Au recovery is achieved even when other base metal ions coexist in the presence of Cu ions.

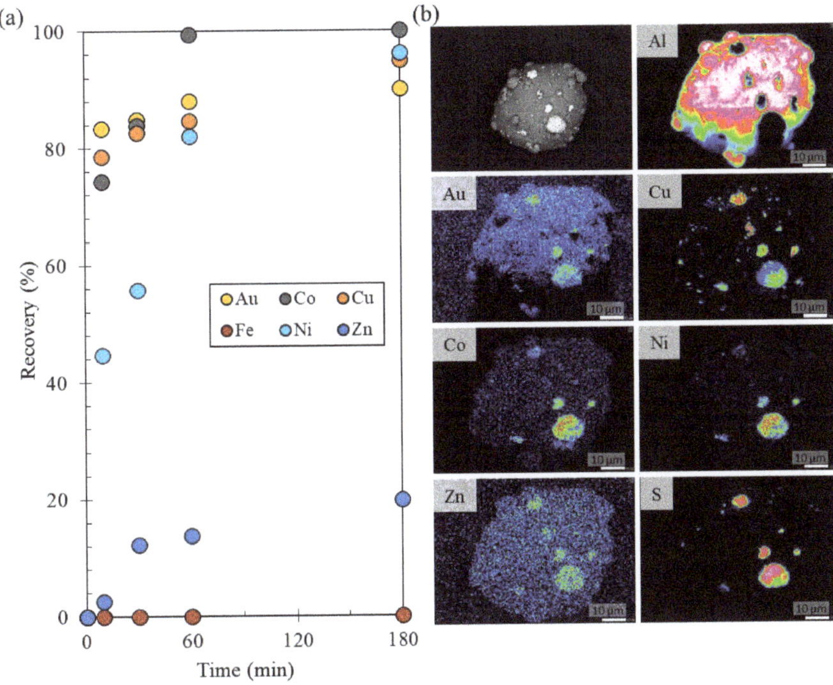

Figure 10. (**a**) Metal ion recovery from the ammonium thiosulfate solution containing all metal ions (i.e., Au, Co, Cu, Fe, Ni, and Zn ions) over time, and (**b**) SEM photomicrographs and corresponding elemental maps of the residue.

The components/concentrations of the elements are different depending on the type as well as the location of the ores. Based on previous studies, typical Cu, Co, Ni, and Zn concentrations in the ore are: Cu at about 250 to 5000 ppm [38,42], Co at about 12 to 50 ppm [39,40], Ni at about 0.002% to 0.03% [43,44], and Zn at about 0.002% to 1.7% [5,43–45] (explanation of Fe content was excluded since it does not affect the current leaching/recovery system). In the absence of Cu ions in the solution, the results showed that Au recovery decreased from 0.03 Co/Al mass ratio and 0.04 Ni/Al mass ratio (in the case of Zn ion addition, Au recovery continuously increased to over 0.22 Zn/Al mass ratio) while high and constant Au recovery was obtained in the presence of Cu ions in the solution regardless of the Me/Al mass ratio under the current conditions. In the ammonium thiosulfate system, Cu ions are essentially employed as a catalyst for enhancing the Au extraction rate by 25-fold, and successful Au recovery was obtained from a solution containing various base metal ions. Note that a much higher concentration of the elements can be present in the ore and affect the recovery system, even in the presence of proper Cu ions. In this case, an increase in ZVAl mass ratio could be expected to increase Au recovery. Thus, this novel recovery technique using galvanic interactions between ZVAl and AC can be employed in the treatment of Au ore containing various elements.

4. Conclusions

The present study describes the applicability of a simple and high-efficiency recovery technique by enhanced cementation of galvanic interaction between ZVAl and AC for Au ore processing, which includes:

- Recovery of Au ions from the solution with low Au concentrations of about less than 10 mg/L;

- Investigation of the effects of various coexisting metal ions that could be present in ore for Au recovery.

The recovery results showed that recovery efficiency slightly decreases as the initial concentration decreases, but still showed over 85% recovery. Selected coexisting Cu, Fe, Co, Ni, and Zn ions were studied to establish their effects on Au recovery, and the results showed that the Au recovery improved (>85–90%) when Cu ions were present in the solution with minimal effects of other competing base metal ions. The recovery of Au ions from the solution containing Cu, Fe, Co, Ni, and Zn ions was also carried out and the results showed that even if various metal ions coexisted together with Au ions in the solution, over 90% of Au could be recovered. In the ammonium thiosulfate system, Cu ions are employed as an essential catalyst; thus, the results showed a high possibility of applicability to Au ore mining processes.

Author Contributions: Conceptualization, S.J., I.P., M.I. and N.H.; methodology, S.J. and N.H.; investigation, S.J. and S.B.; writing—original draft preparation, S.J. and S.B.; writing—review and editing, S.J., I.P., C.B.T., M.I. and N.H.; project administration, S.J., I.P., M.I. and N.H.; funding acquisition, S.J. All authors have read and agreed to the published version of the manuscript.

Funding: This study was financially supported by Japan Oil, Gas and Metals National Corporation (JOGMEC).

Data Availability Statement: Data available on request due to restrictions, as the research is ongoing.

Conflicts of Interest: The authors declare no conflict of interest.

References

1. Aylmore, M.; Muir, D. Thiosulfate leaching of gold—A review. *Miner. Eng.* **2001**, *14*, 135–174. [CrossRef]
2. Tabelin, C.B.; Park, I.; Phengsaart, T.; Jeon, S.; Villacorte-Tabelin, M.; Alonzo, D.; Yoo, K.; Ito, M.; Hiroyoshi, N. Copper and critical metals production from porphyry ores and e-wastes: A review of resource availability, processing/recycling challenges, socio-environmental aspects, and sustainability issues. *Resour. Conserv. Recycl.* **2021**, *170*, 105610. [CrossRef]
3. Opiso, E.M.; Aseneiro, J.P.J.; Banda, M.H.T.; Tabelin, C.B. Solid-phase partitioning of mercury in artisanal gold mine tailings from selected key areas in Mindanao, Philippines, and its implications for mercury detoxification. *Waste Manag. Res.* **2018**, *36*, 269–276. [CrossRef] [PubMed]
4. Jeon, S.; Tabelin, C.B.; Takahashi, H.; Park, I.; Ito, M.; Hiroyoshi, N. Interference of coexisting copper and aluminum on the ammonium thiosulfate leaching of gold from printed circuit boards of waste mobile phones. *Waste Manag.* **2018**, *81*, 148–156. [CrossRef] [PubMed]
5. Aazami, M.; Lapidus, G.; Azadeh, A. The effect of solution parameters on the thiosulfate leaching of Zarshouran refractory gold ore. *Int. J. Miner. Process.* **2014**, *131*, 43–50. [CrossRef]
6. Ha, V.H.; Lee, J.-C.; Huynh, T.H.; Jeong, J.; Pandey, B. Optimizing the thiosulfate leaching of gold from printed circuit boards of discarded mobile phone. *Hydrometallurgy* **2014**, *149*, 118–126. [CrossRef]
7. Jeon, S.; Tabelin, C.B.; Park, I.; Nagata, Y.; Ito, M.; Hiroyoshi, N. Ammonium thiosulfate extraction of gold from printed circuit boards (PCBs) of end-of-life mobile phones and its recovery from pregnant leach solution by cementation. *Hydrometallurgy* **2020**, *191*, 105214. [CrossRef]
8. Molleman, E.; Dreisinger, D. The treatment of copper–gold ores by ammonium thiosulfate leaching. *Hydrometallurgy* **2002**, *66*, 1–21. [CrossRef]
9. Ficeriová, J.; Baláz, P.; Villachica, C.L. Thiosulfate leaching of silver, gold and bismuth from a complex sulfide concentrates. *Hydrometallurgy* **2005**, *77*, 35–39. [CrossRef]
10. Fleming, C.A.; Mezei, A.; Bourricaudy, E.; Canizares, M.; Ashbury, M. Factors influencing the rate of gold cyanide leaching and adsorption on activated carbon, and their impact on the design of CIL and CIP circuits. *Miner. Eng.* **2011**, *24*, 484–494. [CrossRef]
11. Navarro, P.; Vargas, C.; Alonso, M.; Alguacil, F. The adsorption of gold on activated carbon from thiosulfate-ammoniacal solutions. *Gold Bull.* **2006**, *39*, 93–97. [CrossRef]
12. Arima, H.; Fujita, T.; Yen, W.-T. Gold Cementation from Ammonium Thiosulfate Solution by Zinc, Copper and Aluminium Powders. *Mater. Trans.* **2002**, *43*, 485–493. [CrossRef]
13. Hiskey, J.B.; Lee, J. Kinetics of gold cementation on copper in ammoniacal thiosulfate solutions. *Hydrometallurgy* **2003**, *69*, 45–56. [CrossRef]
14. Dong, Z.; Jiang, T.; Xu, B.; Yang, Y.; Li, Q. Recovery of Gold from Pregnant Thiosulfate Solutions by the Resin Adsorption Technique. *Metals* **2017**, *7*, 555. [CrossRef]

15. Jeon, S.; Tabelin, C.B.; Takahashi, H.; Park, I.; Ito, M.; Hiroyoshi, N. Enhanced cementation of gold via galvanic interactions using activated carbon and zero-valent aluminum: A novel approach to recover gold ions from ammonium thiosulfate medium. *Hydrometallurgy* **2020**, *191*, 105165. [CrossRef]
16. Vaughan, J.P. The process mineralogy of gold: The classification of ore types. *JOM* **2004**, *56*, 46–48. [CrossRef]
17. Cho, K.; Kim, H.; Myung, E.; Purev, O.; Choi, N.; Park, C. Recovery of Gold from the Refractory Gold Concentrate Using Microwave Assisted Leaching. *Metal* **2020**, *10*, 571. [CrossRef]
18. Qin, H.; Guo, X.; Tian, Q.; Yu, D.; Zhang, L. Recovery of gold from sulfide refractory gold ore: Oxidation roasting pretreatment and gold extraction. *Miner. Eng.* **2021**, *164*, 106822. [CrossRef]
19. Islam, K.; Vilaysouk, X.; Murakami, S. Integrating remote sensing and life cycle assessment to quantify the environmental impacts of copper-silver-gold mining: A case study from Laos. *Resour. Conserv. Recycl.* **2020**, *154*, 104630. [CrossRef]
20. Yang, Y.; Gao, W.; Xu, B.; Li, Q.; Jiang, T. Study on oxygen pressure thiosulfate leaching of gold without the catalysis of copper and ammonia. *Hydrometallurgy* **2019**, *187*, 71–80. [CrossRef]
21. Gorji, M.; Hosseini, M.R.; Ahmadi, A. Comparison and optimization of the bio-cyanidation potentials of *B. megaterium* and *P. aeruginosa* for extracting gold from an oxidized copper-gold ore in the presence of residual glycine. *Hydrometallurgy* **2020**, *191*, 105218. [CrossRef]
22. Agorhom, E.A.; Owusu, C. The Effects of Pulp Rheology on Gravity Gold Recovery in Free Milling Gold Ore of the Tarkwaian Systems of Ghana. *Miner. Process. Extr. Met. Rev.* **2020**, 1–6. [CrossRef]
23. Tabelin, C.B.; Silwamba, M.; Paglinawan, F.C.; Mondejar, A.J.S.; Duc, H.G.; Resabal, V.J.; Opiso, E.M.; Igarashi, T.; Tomiyama, S.; Ito, M.; et al. Solid-phase partitioning and release-retention mechanisms of copper, lead, zinc and arsenic in soils impacted by artisanal and small-scale gold mining (ASGM) activities. *Chemosphere* **2020**, *260*, 127574. [CrossRef]
24. Park, I.; Hong, S.; Jeon, S.; Ito, M.; Hiroyoshi, N. A Review of Recent Advances in Depression Techniques for Flotation Separation of Cu–Mo Sulfides in Porphyry Copper Deposits. *Metal* **2020**, *10*, 1269. [CrossRef]
25. Park, I.; Higuchi, K.; Tabelin, C.B.; Jeon, S.; Ito, M.; Hiroyoshi, N. Suppression of arsenopyrite oxidation by microencapsulation using ferric-catecholate complexes and phosphate. *Chemosphere* **2021**, *269*, 129413. [CrossRef] [PubMed]
26. Sahoo, P.; Venkatesh, A. Constraints of mineralogical characterization of gold ore: Implication for genesis, controls and evolution of gold from Kundarkocha gold deposit, eastern India. *J. Asian Earth Sci.* **2015**, *97*, 136–149. [CrossRef]
27. Wu, J.; Ahn, J.; Lee, J. Gold deportment and leaching study from a pressure oxidation residue of chalcopyrite concentrate. *Hydrometallurgy* **2021**, *201*, 105583. [CrossRef]
28. Adams, M.; Lawrence, R.; Bratty, M. Biogenic sulphide for cyanide recycle and copper recovery in gold–copper ore processing. *Miner. Eng.* **2008**, *21*, 509–517. [CrossRef]
29. Wang, Z.; Chen, D.; Chen, L. Application of fluoride to enhance aluminum cementation of gold from acidic thiocyanate solution. *Hydrometallurgy* **2007**, *89*, 196–206. [CrossRef]
30. Nguyen, H.; Tran, T.; Wong, P. A kinetic study of the cementation of gold from cyanide solutions onto copper. *Hydrometallurgy* **1997**, *46*, 55–69. [CrossRef]
31. Silwamba, M.; Ito, M.; Hiroyoshi, N.; Tabelin, C.B.; Fukushima, T.; Park, I.; Jeon, S.; Igarashi, T.; Sato, T.; Nyambe, I.; et al. Detoxification of lead-bearing zinc plant leach residues from Kabwe, Zambia by coupled extraction-cementation method. *J. Environ. Chem. Eng.* **2020**, *8*, 104197. [CrossRef]
32. Arima, H.; Fujita, T.; Yen, W.-T. Using Nickel as a Catalyst in Ammonium Thiosulfate Leaching for Gold Extraction. *Mater. Trans.* **2004**, *45*, 516–526. [CrossRef]
33. Liu, X.; Jiang, T.; Xu, B.; Zhang, Y.; Li, Q.; Yang, Y.; He, Y. Thiosulphate leaching of gold in Cu-NH3-S2O32−—H2O system: An updated thermodynamic analysis using predominance area and species distribution diagrams. *Miner. Eng.* **2020**, *151*, 106336. [CrossRef]
34. Liu, X.; Xu, B.; Yang, Y.; Li, Q.; Jiang, T.; He, Y. Thermodynamic analysis of ammoniacal thiosulfate leaching of gold catalyzed by Co(III)/Co(II) using Eh-pH and speciation diagrams. *Hydrometallurgy* **2018**, *178*, 240–249. [CrossRef]
35. Xu, B.; Li, K.; Li, Q.; Yang, Y.; Liu, X.; Jiang, T. Kinetic studies of gold leaching from a gold concentrate calcine by thiosulfate with cobalt-ammonia catalysis and gold recovery by resin adsorption from its pregnant solution. *Sep. Purif. Technol.* **2019**, *213*, 368–377. [CrossRef]
36. BC Campus, Appendix: Standard Reduction Potentials by Value. Available online: https://opentextbc.ca/introductorychemistry/back-matter/appendix-standard-reduction-potentials-by-value-2/ (accessed on 1 April 2021).
37. Senanayake, G.; Senaputra, A.; Nicol, M.J. Effect of thiosulfate, sulfide, copper(II), cobalt(II)/(III) and iron oxides on the ammoniacal carbonate leaching of nickel and ferronickel in the caron process. *Hydrometallurgy* **2010**, *105*, 60–68. [CrossRef]
38. Djokic, S.S. *Electroless Depositiore on of Metals and Alloys, Modern Aspects of Electrochemistry*; Brian, E.C., Conway, E., Ralph, E.W., White, E., Eds.; Springer: Berlin/Heidelberg, Germany, 2002; pp. 51–133.
39. Senanayake, G.; Zhang, X.M. Gold leaching by copper (II) in ammoniacal thiosulfate solutions in the presence of additives. Part II: Effects of residual Cu(II), pH and redox potentials on reactivity of colloidal gold. *Hydrometallurgy* **2012**, *115-116*, 21–29. [CrossRef]
40. Compare Metals. Available online: https://metals.comparenature.com/en/copper-vs-cobalt/comparison-6-30-0 (accessed on 15 April 2021).
41. New Medical Device Shielding Requirements and Die Casting (IEC 60601-0-2: 2014—4th Edition). Available online: https://www.abdiecasting.com/new-medical-device-shielding-requirements-and-die-casting/ (accessed on 26 May 2021).

42. Örgül, S.; Atalay, Ü. Reaction chemistry of gold leaching in thiourea solution for a Turkish gold ore. *Hydrometallurgy* **2002**, *67*, 71–77. [CrossRef]
43. Soltani, F.; Darabi, H.; Badri, R.; Zamankhan, P. Improved recovery of a low-grade refractory gold ore using flotation-preoxidation-cyanidation methods. *Int. J. Min. Sci. Technol.* **2014**, *24*, 537–542.
44. Melashvili, M.; Fleming, C.; Dymov, I.; Matthews, D.; Dreisinger, D. Dissolution of gold during pyrite oxidation reaction. *Miner. Eng.* **2016**, *87*, 2–9. [CrossRef]
45. Murthy, D.S.R.; Kumar, V.; Rao, K.V. Extraction of gold from an Indian low-grade refractor gold ore through physical beneficiation and thiourea leaching. *Hydrometallurgy* **2003**, *68*, 125–130. [CrossRef]

Article

Development of Hydrometallurgical Process for Recovery of Rare Earth Metals (Nd, Pr, and Dy) from Nd-Fe-B Magnets

Pankaj Kumar Choubey [1], Nityanand Singh [1], Rekha Panda [1], Rajesh Kumar Jyothi [2], Kyoungkeun Yoo [3], Ilhwan Park [4],* and Manis Kumar Jha [1],*

1   Metal Extraction and Recycling Division, CSIR-National Metallurgical Laboratory, Jamshedpur 831007, India; impankaj.choubey@gmail.com (P.K.C.); nityanandsinghjsr@gmail.com (N.S.); rekhapanda1608@gmail.com (R.P.)
2   Mineral Resources Research Division, Korea Institute of Geoscience and Mineral Resources, Daejeon 34132, Korea; rkumarphd@kigam.re.kr
3   Department of Energy and Resources Engineering, Korea Maritime and Ocean University (KMOU), Busan 49112, Korea; kyoo@kmou.ac.kr
4   Division of Sustainable Resources Engineering, Faculty of Engineering, Hokkaido University, Sapporo 060-8606, Japan
*   Correspondence: i-park@eng.hokudai.ac.jp (I.P.); mkjha@nmlindia.org (M.K.J.)

Abstract: Non-availability of rich primary resources of rare earth metals (REMs) and the generation of huge amounts of discarded magnets containing REMs, compelled the researchers to explore the possibilities for the recovery of REMs from discarded magnets. Therefore, the present paper reports the recovery of REMs (Nd, Pr, and Dy) from discarded Nd-Fe-B magnets. The process consists of demagnetization, pre-treatment, and hydrometallurgical processing to recover REMs as salt. Leaching studies indicate that 95.5% Nd, 99.9% Pr, and 99.9% Dy were found to be dissolved at the optimized experimental condition i.e., acid concentration 2 M H_2SO_4, temperature 75 °C, pulp density 100 g/L, and mixing time 60 min. Solvent extraction technique was tried for the selective extraction/separation of REMs and Fe. The result indicates that 99.1% (24.42 g/L) of Nd along with 90% (1.08 g/L) of Pr and total Fe were co-extracted using 35% Cyanex 272 at organic to aqueous (O/A) ratio 1/1, eq. pH 3.5 in 10 min of mixing time. It requires multistage separation and therefore, not feasible in view of economics. Thus, direct precipitation of REMs salt and iron oxide as pigment was studied using two stages of precipitation at different pH. The obtained precipitate of REMs and Fe hydroxides were dried separately to remove the moisture and further treated at elevated temperature to get pure REMs oxide and red oxide.

Keywords: REMs; secondary resources; recycling; pretreatment; hydrometallurgy; Nd-Fe-B magnet

Citation: Choubey, P.K.; Singh, N.; Panda, R.; Jyothi, R.K.; Yoo, K.; Park, I.; Jha, M.K. Development of Hydrometallurgical Process for Recovery of Rare Earth Metals (Nd, Pr, and Dy) from Nd-Fe-B Magnets. *Metals* **2021**, *11*, 1987. https://doi.org/10.3390/met11121987

Academic Editor: Alberto Moreira Jorge Junior

Received: 2 November 2021
Accepted: 6 December 2021
Published: 9 December 2021

1. Introduction

Due to rapid industrialization and technological advancement, old electronic devices are replaced with new ones, resulting in the generation of massive amounts of discarded electronic scraps such as personal computers, mobile phones, televisions, etc. [1]. It has been reported that 53.5 million tons (Mt) of e-waste were generated in 2019 and it has been expected that 74.7 Mt of e-waste will be generated by 2030 [2]. The discarded e-waste contains a variety of metals such as base metals (Cu, Ni, Sn, Pb, etc.), rare earth metals (Nd, Pr, Dy, etc.) including hazardous metals that may cause a serious environmental threat if disposed into the landfilling without treatment, as well as the loss of valuables. Therefore, it is needed to develop a sustainable process for the extraction of metals from e-waste, which may also minimize the dependency on primary natural resources.

Various studies have been made to recover rare earth metals (REMs) using pyro-/hydro-metallurgy and hybrid processes. Onal et al. [3] reported the nitration of Nd-Fe-B magnet at 200 °C to convert the refractory magnet into soluble species of REMs. Further, calcined magnets were leached in water at room temperature to dissolve more than 95% REMs while maintaining a pulp density of 60 g/L. In addition, a magnet was roasted at 800 °C to transform the refractory magnet into soluble species of REMs. The roasted product was pressure leached in 0.6 M hydrochloric acid at 180 °C to leach 99% REMs [4]. The mechano-chemical studies have also been made with the addition of ferric sulfate to convert the refractory magnet into their sulfate salt to make the subsequent leaching step easier. The pre-treated product was further leached in water at elevated temperature and subsequently, REMs were precipitated with oxalic acid at pH 1.9 [5].

In addition, oxidative roasting followed by organic acid leaching has been used to recover the REMs from Nd-Fe-B magnet. The refractory Nd-Fe-B magnet was oxidized into the acid susceptible species by heating at 900 °C for 480 min. The oxidized product was dissolved in a mixture of malic and citric acids at 90 °C to leach more than 90% REMs [6]. Kumari et al. [1] reported the sulfuric acid leaching and precipitation process for recovery of Nd, almost ~99.99% REMs (Nd, Dy and Pr) were leached in 1 M sulfuric acid at room temperature. Further, Nd was precipitated with ~98% purity by using ammonia to get Nd salt at pH 1.65; however, minor co-precipitation of Dy and Pr also occurred. In a subsequent study, electrochemical leaching was conducted in a mixture of 0.1 M H_2SO_4 and 0.05 M $H_2C_2O_4$ at the current density of 20 A/dm^2 to enhance the leaching efficiency of REMs [7]. The obtained leach liquor was subjected to separation and purification studies to extract the REMs. Pavon et al. [8] extracted 99.9% Nd from the leach liquor of Nd-Fe-B magnet at eq. pH 0.8 to 1.5 with 0.3 M Cyanex 572. Further, 58.62% Nd, 98.46% Dy, and 85.59% Pr were extracted at pH 2 with 1 M D2EHPA and stripped with 2 M nitric acid [9].

Besides the solvent extraction, precipitation studies have also been investigated for the extraction of REMs from leach liquor of Nd-Fe-B magnet. Lee et al. [10] used sodium hydroxide to precipitate Nd as $Nd(OH)_3$ at pH 0.6. Rabatho et al. [11] leached 98% of Nd and 81% of Dy in 1 M HNO_3 at 25 °C in presence of 0.3 M H_2O_2, but Fe remained in the residue. Further, 81.8% Dy and 91.5% Nd were recovered by precipitation using oxalic acid ($H_2C_2O_4$) in a range of pH 8 at room temperature [11]. A lot of research on this topic is reported, but due to various factors it is still not applied for commercial exploitation in industries. Another aspect of this research is recycling, which will play an important role in the circular economy as well as in the conservation of natural resources. The important findings in the developed process are closed-loop, which means all the materials will be recycled in the production flow with no waste, i.e., zero waste generation.

Keeping in view of the above, the present paper reports the novel process for selective recovery of REMs oxides (Nd, Pr, and Dy) from discarded magnets of the hard disk considering the cost of the process. The magnets were separated by being dismantled, demagnetized by heat treatment, crushed to reduce the particle size, leached to dissolve the REMs, and selective extraction of REMs using solvent extraction/precipitation to produce REMs oxides. Compared to previous studies, the developed process reports less energy for demagnetization of discarded Nd-Fe-B magnets as well as a smaller number of stages to get REMs like rare earth oxides.

2. Materials and Methods

2.1. Pre-Treatment of Nd-Fe-B Magnets

At first, a discarded hard disk of CPUs was dismantled to separate the Nd-Fe-B magnets for recovery of rare earth metals including iron. Further, Nd-Fe-B magnets were demagnetized by heating at 300 °C for 3 h. The demagnetized material was crushed into small sizes (2 × 2 mm) using the mortar pestle to make the particle surface area higher, which will enable effective leaching. The process flow-sheet for the pre-treatment of Nd-Fe-B magnet is presented as Figure 1. The composition of Nd-Fe-B magnet was determined by the chemical analysis method and presented in Table 1.

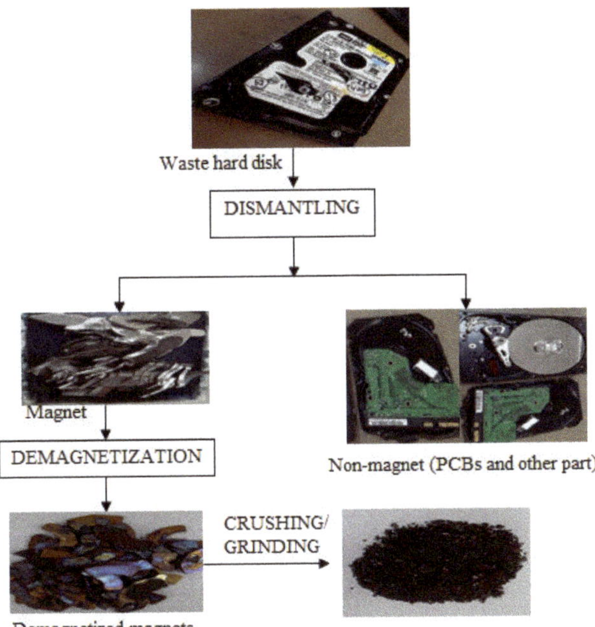

Figure 1. Pre-treatment of magnets prior to leaching of REMs.

Table 1. Composition of metals present in Nd-Fe-B magnet.

REMs	Contents (wt %)	Non-REMs	Contents (wt %)
Nd	29.5	Fe	59.5
Dy	2.5	Ni	4.1
Pr	1.3	Co	3.1

2.2. Leaching Procedure

Dissolution experiments were carried out in a leaching reactor of capacity: 100 mL (Borosil, Mumbai, India) equipped with a condenser facility to avoid the loss of liquid through evaporation due to heating. Leaching studies were carried out using a hot plate fitted with a temperature controller and sensor. To optimize the process parameters for the effective dissolution of REMs, the concentration of sulfuric acid leachant and temperature were varied between 0.5 to 2.5 M and 25 to 90 °C, respectively. To study the dissolution behavior of REMs along with other metals, the samples of slurry/leach liquor were collected at regular intervals of time during the leaching experiment. The leach liquor was separated from the leached residue using filtration method. The metals present in the leached residue and leach liquor were checked to see the material balance. Satisfactory material balance was obtained for each set of leaching experiments by calculating the weight of the sample before leaching, which is equal to the amount of metals concentration in solution plus the weight of leached residue. Further, the leach liquor was processed for REMs extraction using the solvent extraction technique.

2.3. Solvent Extraction Procedure

Solvent extraction studies were carried out in a conical flask using a magnetic stirrer (Borosil, Mumbai, India) at room temperature. The equal volume of leach liquor and extractant Cyanex 272 (Figure 2) (50 mL/50 mL) were put in the conical flask for proper mixing. Ammonium hydroxide was used to maintain the eq. pH in the range of 1 to

3.5. When the solution attained the equilibrium, the metal-loaded organic extractant and raffinate were separated using a separating funnel. The metals-loaded organic extractant was stripped using dil. sulfuric acid. The concentration of metal ions in the loaded extractant and raffinate were analyzed by using ICP-OES (VISTA–PMX, CCD Simultaneous, Make: Australia). The material balance was checked by calculating the metal ions present in the strip solution obtained from loaded extractant, raffinate, and the head sample.

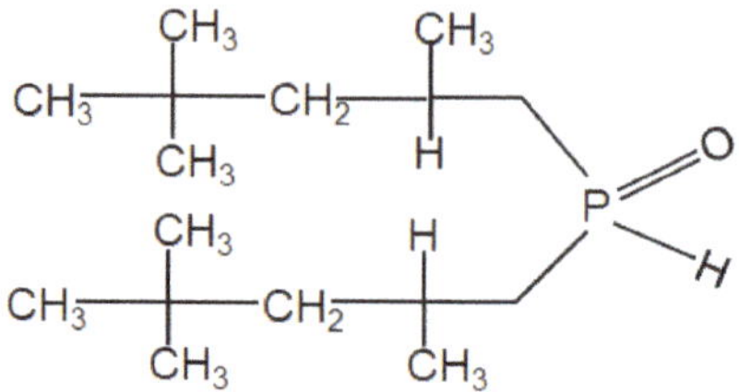

Figure 2. Structure of Cyanex 272 [bis (2,2,4 trimethylpentyl) phosphinic acid].

2.4. Precipitation Procedure

Precipitation experiments were conducted in a beaker (capacity: 100 mL) under the constant stirring speed (300 rpm) at standard temperature using ammonium hydroxide as a precipitant. The solution pH was varied in the range of 1.0 to 2.5 by the addition of ammonium hydroxide (25.96 M) for the precipitation of REMs. During the precipitation reaction, samples were collected at regular intervals of time to observe the precipitation behavior of metals at various pH and time.

2.5. Characterization and Analysis of Samples

Non-ferrous metals were analyzed using Atomic Absorption Spectrophotometer, AAS, Analyst 200 (Perkin Elmer, Waltham, MA, USA), and REMs were analyzed using Inductively Coupled Plasma-Optical Emission Spectrophotometer, ICP-OES, VISTA-PMX, CCD Simultaneous (Australia). X-ray diffraction (XRD, Bruker AXS D8 instrument) (Bruker, Wisconsin, WI, USA) and scanning electron microscope energy dispersive X-ray spectroscopy (SEM-EDS, JXA-8230 Electron Probe Micro Analyzer, JEOL) (JEOL, Tokyo, Japan) were carried to analyze the phases as well as morphological appearances of the beneficiated magnet. XRD pattern shows that Nd-Fe-B magnet mainly contains the peak of $Nd_2Fe_{14}B$ (Figure 3) and was also confirmed by EDS analysis (Figure 4). The Eh- pH of the solution was analyzed (Eh: 0.55 to 1.05 V) using Eutech pH 700 (Thermo Fisher Scientific, 7 Gul Circle, level 2M, Keppel Logistic Building, Singapore). Analytical grade (AR) ammonium hydroxide, sulfuric acid, hydrochloric acid supplied by Rankem, India, were used for experimental purposes.

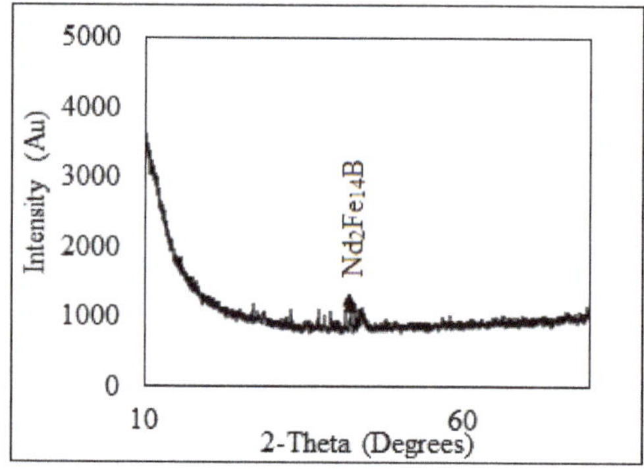

Figure 3. XRD pattern of crushed magnet.

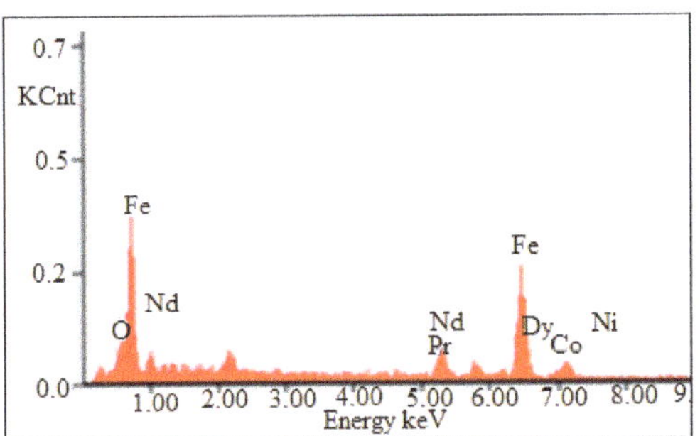

Figure 4. EDS pattern of demagnetized crushed Nd-Fe-B magnet.

3. Results and Discussion

In order to recover the REMs from discarded Nd-Fe-B magnets, leaching, solvent extraction, and precipitation experiments were carried out. At first, Nd-Fe-B magnets were demagnetized by heat treatment. The demagnetized material was crushed to get fine powder for making it suitable for the effective leaching of REMs. The leaching studies were conducted varying different experimental parameters to get suitable conditions for metals dissolution. Further, from the leach liquor, solvent extraction studies were made to extract the REMs, particularly, Nd from the leach liquor. The loaded organic was stripped to get Nd enriched solution. Finally, the precipitation studies were carried out to recover the Nd as precipitate with other elements as minor/trace. The obtained results are discussed below.

3.1. Leaching Study

For the effective dissolution of metals from waste Nd-Fe-B magnets, the various experimental parameters, i.e., acid concentration, reaction time, temperature, pulp density, etc. were studied to leach the REMs effectively from the demagnetized and crushed powder of magnets.

3.1.1. Effect of Acid Concentration

Leaching experiments were performed using different acid concentrations of leachant varied from 0.5 to 2.5 M at temp. 60 °C and mixing time 60 min maintaining pulp density 100 g/L. Results (Figure 5) show that the dissolution of REMs (Nd, Pr and Dy) was found to be increased with increasing acid concentration due to the increase in the acidic strength of the leachant. But, above the acid concentration of 2 M, the leaching efficiency of REMs remained the same. This indicates that 2 M H_2SO_4 has enough acidity to leach the REMs. Hence, 2 M H_2SO_4 was chosen as the optimum concentration of acid for the dissolution of REMs from spent Nd-Fe-B magnets. The chemical reaction involved during the leaching of REMs is presented in Equation (1):

$$2Nd_2Fe_{14}B + 37H_2SO_4 \rightarrow 2Nd_2(SO_4)_3 + 28FeSO_4 + B_2(SO_4)_3 + 18.5H_2 \tag{1}$$

3.1.2. Effect of Temperature

To optimize the temperature for leaching of REMs, experiments were performed at various experimental temperatures varying in the range of 60 to 90 °C at a pulp density of 100 g/L. Figure 6 showed that leaching efficiency of Nd, Pr, and Dy enhanced with the increase in solution temperature due to the increase in the rate of reaction. A total of ~95.5% Nd was found to be leached at 75 °C and a further increase in solution temperature above 75 °C had no significant enhancement on the dissolution of REMs. Therefore, 75 °C was chosen as the optimum temperature for leaching of REMs for further sets of leaching experiments.

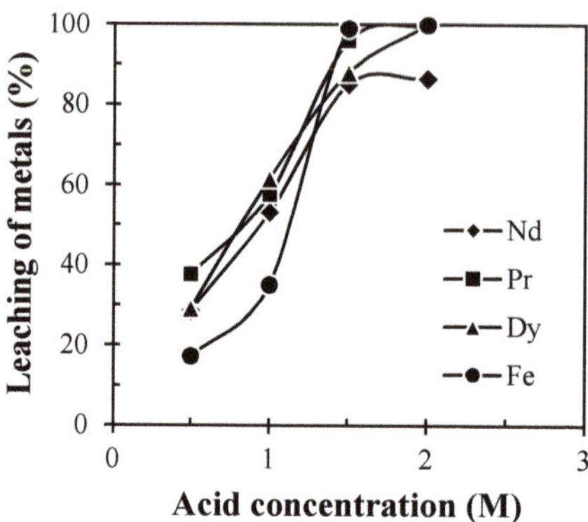

Figure 5. Effect of acid concentration on leaching of REMs from Nd-Fe-B magnets (Solid: Crushed magnet; Liquid: 0.5 to 2 M H_2SO_4; Pulp density: 100 g/L; Temp.: 60 °C; Time: 60 min).

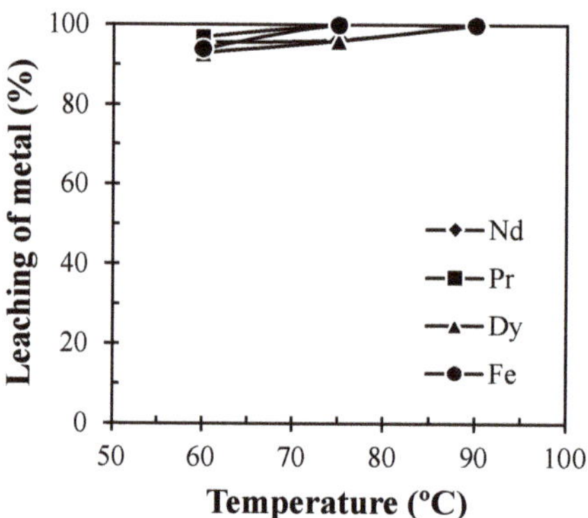

Figure 6. Effect of temperature on leaching of REMs from Nd-Fe-B magnets (Solid: Crushed magnet; Liquid: 2 M H_2SO_4; Pulp density: 100 g/L; Time: 60 min).

3.1.3. Effect of Pulp Density

For the optimization of solid to liquid ratio, experiments were performed in a range of pulp density variations from 25 to 200 g/L at 60 °C in 60 min. It was found that leaching of REMs remains the same in the range of pulp density 25 to 100 g/L (Figure 7), but leaching efficiency of REMs was found to be decreased at the pulp density above 100 g/L. This indicates that the number of moles of metallic constitutes (Nd, Pr, and Dy) of REMs became higher than that of leachant molecules, resulting in the decrease in the leaching efficiency of REMs above the pulp density of 100 g/L. Therefore, 100 g/L pulp density was chosen as the optimum condition for the dissolution of REMs from the discarded magnet.

3.2. Solvent Extraction of Nd and Pr

Selective extraction and stripping of REMs/Fe using solvent extraction technique were tried. Experiments were carried out to recover Nd from the obtained leach liquor containing 24.45 g/L Nd, 1.2 g/L Pr, 2.49 g/L Dy, and 49.76 g/L Fe using Cyanex 272. It was found that extraction of Nd increases with the increase in eq. pH, while Pr and Fe also co-extracted (Figure 8). About 15.59 g/L Nd were extracted using 35% Cyanex 272 at eq. pH 0.75 in 10 min mixing time, while complete extraction of Nd occurred at eq. pH 3. Further increase in eq. pH had no significant effect on the extraction of Nd. Further increase in eq. pH had no significant effect on the extraction of Nd. Hence, eq. pH 3 was considered as optimum pH for the extraction of Nd from leach liquor. Further, extraction co-efficient (Kd) (Equation (2)) and separation factor ($\beta_{Nd/Pr}$) (Equation (3)) of Nd were calculated and found to be 1.5 and 4.42, respectively. The low value of separation factor and extraction co-efficient for Nd indicates that the separation can be possible using multistage extraction. The pH was adjusted during the solvent extraction experiment. The pH was not adjusted separately for aqueous feed, therefore, no precipitation occurred. As per the Eh-pH diagram, the REMs (Nd, Pr) were precipitated as their hydroxides at or above pH 1.5. However, at the same pH, REMs form their complexes and had a tendency to be transferred into the organic phase during solvent extraction. The solvent extraction reaction is very fast in comparison to the slow precipitation reaction. Hence, no precipitation was observed during the solvent extraction of REMs under the studied conditions.

$$Kd = \frac{Conc.\ of\ Nd\ in\ organic\ phase}{Conc.\ of\ Nd\ in\ aqueous\ phase} \tag{2}$$

$$\beta = \frac{Extraction\ coefficient\ of\ metal\ Nd}{Extraction\ coefficient\ of\ metal\ Pr} \tag{3}$$

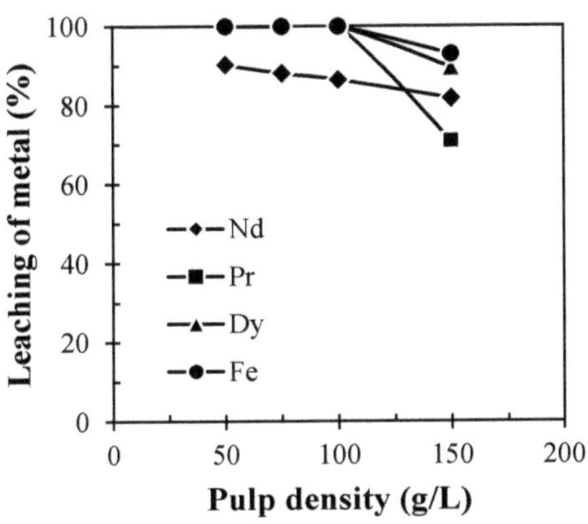

Figure 7. Effect of pulp density on leaching of REMs from Nd-Fe-B magnets (Solid: Crushed magnet; Acid concentration: 2 M H_2SO_4; Temp.: 60 °C; Time: 60 min).

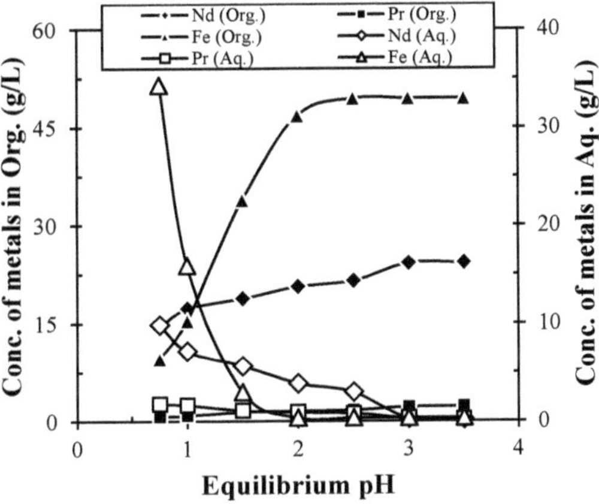

Figure 8. Extraction of Nd and Pr from leach liquor of spent Nd-Fe-B magnet (Extractant: 35% Cyanex 272; Solution: Leach liquor containing 24.45 g/L Nd, 1.2 g/L Pr, 2.49 g/L Dy, and 49.76 g/L Fe; Time: 10 min; O/A ratio: 1/1).

The multistage solvent extraction process for the separation of REMs and Fe will increase the production cost. Therefore, to make the process feasible and economical, direct precipitation studies were carried out.

3.3. Recovery of REMs from Leach Liquor by Precipitation

Further, the precipitation studies were carried out to recover REMs from the obtained leach liquor 24.45 g/L Nd, 1.2 g/L Pr, 2.49 g/L Dy and 49.76 g/L Fe using the precipitation technique. The pH of the solution was varied in the range of 0.5 and 2.0 using ammonium hydroxide to precipitate selectively the REMs at room temperature, where REEs represents the rare earth elements (Equations (4) and (5)):

$$REE_2(SO_4)_3 + 6NH_4OH \rightarrow 2REE(OH)_3 + 3(NH_4)_2SO_4 \tag{4}$$

$$2REE(OH)_3 \rightarrow REE_2O_3 + 3H_2O \tag{5}$$

The precipitation of Nd, Pr, and Dy (Figure 9) was found to be increased with the increase in pH from 0.5 to 2.0 due to the low solubility of REMs at eq. pH 2. Eh-pH diagram drawn using the HSC chemistry version 6.0, which shows that Nd exists as $Nd(OH)_3$ above pH 0.5 (Figure 10), while iron hydroxide formed at pH above 1.5 (Figure 11). Therefore, the pH of the solution was kept between 0.5 to 2 to avoid the co-precipitation of iron during the recovery of REMs hydroxides. The precipitated product was characterized by SEM-EDS (Figure 12), which shows that a strong peak of Nd appeared for the precipitated products. The composition of the precipitated product is shown in Table 2. It confirmed the formation of Nd product in the precipitated sample leaving iron in the solution.

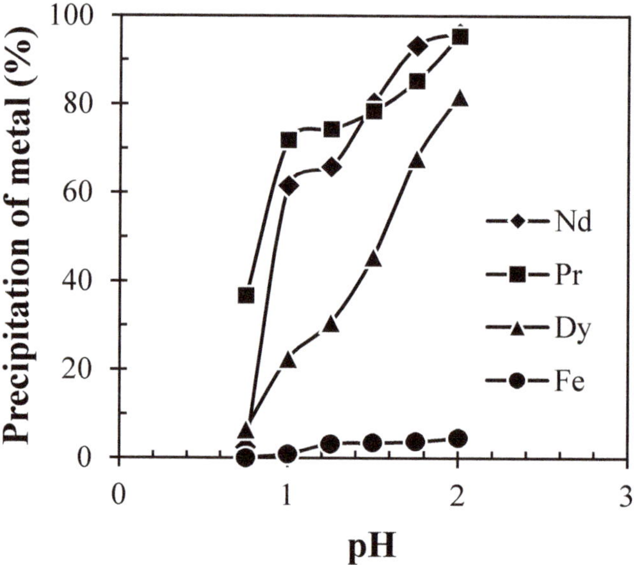

Figure 9. Effect of pH on precipitation of REMs from the liquor of Nd-Fe-B magnets (Solution: Leach liquor containing 24.45 g/L Nd, 1.2 g/L Pr, 2.49 g/L Dy and 49.76 g/L Fe; Precipitant: Dil. Ammonium hydroxide; Time: 15 min).

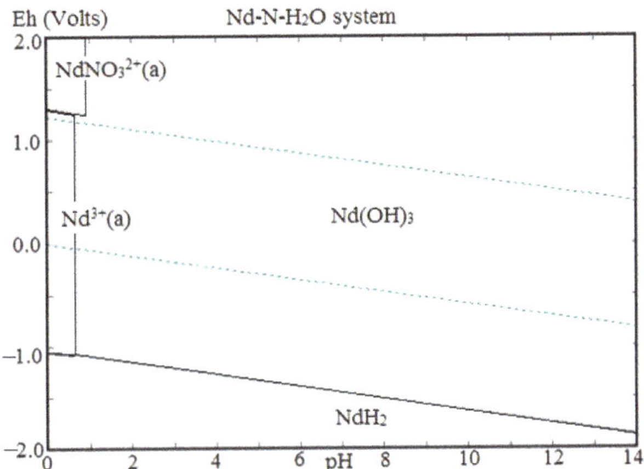

Figure 10. Eh-pH diagram of neodymium.

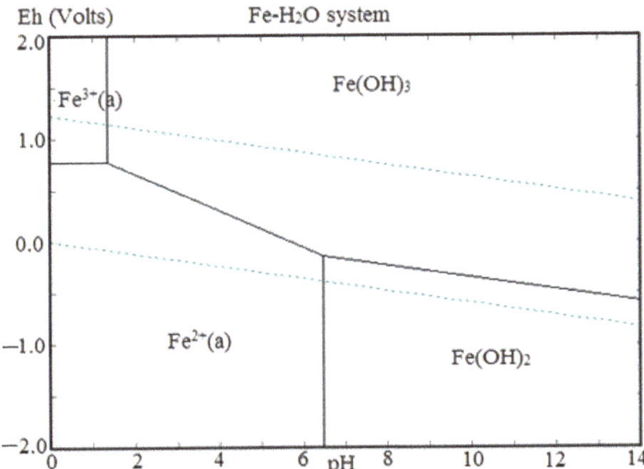

Figure 11. Eh-pH diagram of iron.

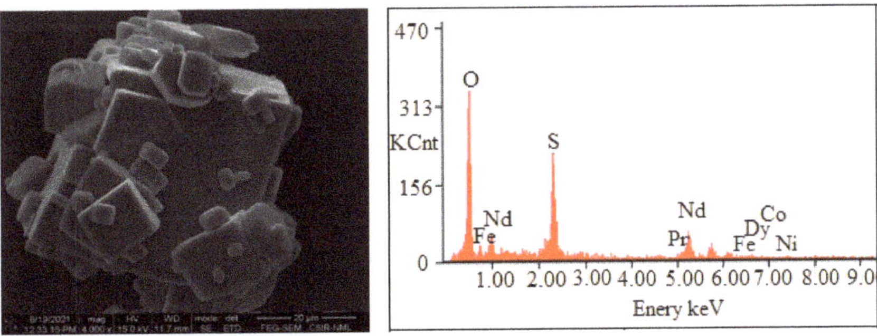

Figure 12. SEM-EDS of precipitated salts of REMs.

Table 2. Composition (EDS) of precipitated salts of REMs.

Element	Wt %	At %
O K	26.1	62.9
S K	16.8	20.2
Pr L	6.10	1.67
Nd L	41.6	11.1
Fe K	1.02	0.70
Dy L	5.03	1.20
Co K	0.91	0.65
Ni K	0.47	0.23

After the separation of REMs, iron (49.7 g) was completely removed by maintaining the pH above 3 using ammonium hydroxide as a precipitant at 60 °C as shown in Equations (6) and (7). It has been assumed that iron was precipitated as Fe^{3+} due to the high stability of ferric ion. The complete process flow sheet has been developed for the dissolution/extraction of REMs from waste Nd-Fe-B magnets of hard disks as shown in Figure 13. Further, comparative data for extraction of REMs from discarded Nd-Fe-B magnets are summarized in Table 3, which reflects that high temperature is required for pre-treatment (demagnetization, roasting) of magnets prior to hydrometallurgical extraction of REMs. However, demagnetization temperature is comparatively low in the present study, which makes the process feasible from an energy point of view.

$$Fe_2(SO_4)_3 + 6NH_4OH \rightarrow 2Fe(OH)_3 + 3(NH_4)_2SO_4 \tag{6}$$

$$2Fe(OH)_3 \rightarrow Fe_2O_3 + 3H_2O \tag{7}$$

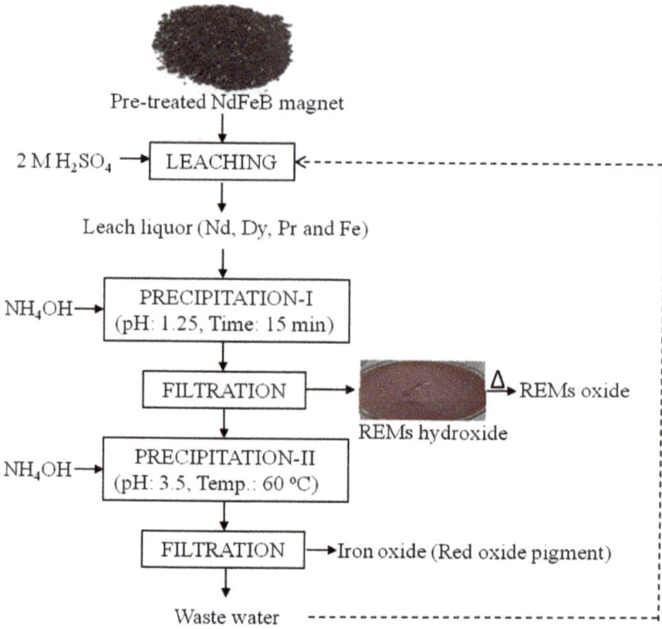

Figure 13. Developed process flow-sheet to recover REMs from spent Nd-Fe-B magnets.

Table 3. Literature review on the extraction of rare earth elements from discarded Nd-Fe-B magnets.

Demagnetization and Roasting Conditions	Experimental Conditions (Leaching and Precipitation/Evaporation)	Salient Results	References
Demagnetization and roasting not done	Leachant: 2 M H_2SO_4, Temp.: 25 °C, Time: 24 h, Evaporation: Temp.: 110 °C	~95% REEs were extracted	Yingnakorn et al. [12]
Demagnetization: 500 °C, Time: 60 min	Not done	Demagnetized	Lee et al. [10]
Demagnetization: 400 °C, Time: 120 min	Not done	Demagnetized	Feng et al. [13]
Demagnetization: 350, Time: 60 min, Roasting: 800 °C, Time: 120 min	Temp.: 180 °C, Time: 120 min; Acid concentration: 0.6 M HCl, Pulp density: 100 g/L Precipitation: Temp.: 50 °C, Initial pH: 2.2, Time: 30 min	99% of REEs were extracted	Liu et al. [4]
Demagnetization: 350 °C, Roasting: 900 °C	Leachant: 1 M Citric/Maleic Acid, Temp.: 90 °C, Pulp density: 50 g/L	99% recovery of Nd occurred	Reisdorfer et al. [6]
Roasting: 950 °C	Leachant: 1.8 M HCl + 3.5 M NH4Cl, Pulp density: 100 g/L, Temp.: 100 °C, Time: 5 days Precipitant: Oxalic acid	90% REMs were dissolved	Hoogerstraete et al. [14]
Roasting: 950 °C	Leachant: 5 M HNO_3, Pulp density: 1000 g/L, Temp.: 80 °C, Time: 72 h Precipitant: Ammonium nitrate, Eq. pH: 2, Time: 60 min, Temp.: 70 °C	Nd and Dy recovered as Nd_2O_3 and Dy_2O_3 with purity 99.6% and 99.8%, respectively	Riano and Binnemans [15]
Roasting: 750 °C	Leachant: Water, Pulp density: 20 g/L, Time:1 h, Temp.: 25 °C	95–100% REMs were extracted	Onal et al. [16]

4. Conclusions

Based on the laboratory scale studies to recover REMs from discarded Nd-Fe-B magnet of hard disks, the following conclusions have been made and are discussed below.

1. It was found that the magnetic strength of Nd-Fe-B hard disk magnet was lost by heat treatment at 300 °C in 2 h.
2. The complete leaching of Nd, Pr, Dy, and Fe occurred from the demagnetized magnet using 2 M H_2SO_4 at 75 °C in 60 min and pulp density 100 g/L.
3. It was found that 99.1% Nd, 94% Pr, and 99.9% Fe got extracted from leach liquor using 35% Cyanex 272 at O/A ratio 1/1 in 10 min. The selective extraction of REMs can only be achieved by using multistage solvent extraction. Thus, direct precipitation studies were carried out at different eq. pH for the selective extraction of REMs and Fe.
4. Further, ~93.15% Nd, ~85% Pr, and ~68% Dy were precipitated from the leach liquor of discarded magnet at pH 1.75 at room temperature in 15 min. The precipitated hydroxide of REMs was converted to their oxides by heating at 120 °C for 2 h.
5. Finally, 96.5% Fe was precipitated as ferric ion through the air sparging between eq. pH 3.5 to 4 at 60 °C.

Author Contributions: Methodology, N.S. and P.K.C.; Validation, P.K.C. and R.P.; Writing—original draft preparation, N.S. and P.K.C.; Writing—review and editing, M.K.J., K.Y. and I.P.; Supervision, M.K.J. and R.K.J.; Project administration, M.K.J. and I.P.; Funding acquisition, I.P. All authors have read and agreed to the published version of the manuscript.

Funding: This study was partially supported by the Japan Society for the Promotion of Science (JSPS) Grant-in-Aid for Early-Career Scientists (JP20K15214). The work has been supported by CSIR-NML under Urban Ore Recycling Centre (FTC-0014/MLP-3116) and various Indo-Korean long term collaboration programs.

Institutional Review Board Statement: Not applicable.

Informed Consent Statement: Not applicable.

Data Availability Statement: The data presented in this study are available on request from the corresponding author.

Acknowledgments: Authors are grateful to the Director CSIR-National Metallurgical Laboratory, Jamshedpur, India, for giving the permission to publish the paper and work done under CSIR-NML projects and international collaborations.

Conflicts of Interest: The authors declare no conflict of interest.

References

1. Kumari, A.; Jha, M.K.; Pathak, D.D. An innovative environmental process for the treatment of scrap Nd-Fe-B magnets. *J. Environ. Manag.* **2020**, *273*, 111063. [CrossRef] [PubMed]
2. Dhir, A.; Malodia, S.; Awan, U.; Sakashita, M.; Kaur, P. Extended valence theory perspective on consumers' e-waste recycling intentions in Japan. *J. Clean. Prod.* **2021**, *312*, 127443. [CrossRef]
3. Onal, M.A.R.; Aktan, E.; Borra, C.R.; Blanpain, B.; Gerven, T.V.; Guo, M. Recycling of NdFeB magnets using nitration, calcination and water leaching for REE recovery. *Hydrometallurgy* **2017**, *167*, 115–123. [CrossRef]
4. Liu, F.; Porvali, A.; Wang, J.; Wang, H.; Peng, C.; Wilson, B.P.; Lundstrom, M. Recovery and separation of rare earths and boron from spent Nd-Fe-B Magnets. *Miner. Eng.* **2020**, *145*, 106097. [CrossRef]
5. Loy, S.V.; Onal, M.A.R.; Binnemans, K.; Gerven, T.V. Recovery of valuable metals from NdFeB magnets by mechanochemically assisted ferric sulfate leaching. *Hydrometallurgy* **2020**, *191*, 105154.
6. Reisdorfer, G.; Bertuol, D.; Tanabe, E.H. Recovery of neodymium from the magnets of hard disk drives using organic acids. *Miner. Eng.* **2019**, *143*, 105938. [CrossRef]
7. Makarova, I.; Soboleva, E.; Osipenko, M.; Kurilo, I.; Laatikainen, M. Electrochemical leaching of rare-earth elements from spent NdFeB magnets. *Hydrometallurgy* **2020**, *192*, 105264. [CrossRef]
8. Pavon, S.; Fortunya, A.; Collb, M.T.; Sastrec, A.M. Neodymium recovery from NdFeB magnet wastes using Primene 81R·Cyanex 572 IL by solvent extraction. *J. Environ. Manag.* **2018**, *222*, 359–367. [CrossRef] [PubMed]
9. Niam, A.C.; Wang, Y.; Chen, S.; Chang, G.; You, S. Simultaneous recovery of rare earth elements from waste permanent magnet (WPMs) leach liquor by solvent extraction and hollow fiber supported liquid membrane. *Chem. Eng. Process.* **2020**, *148*, 107831. [CrossRef]
10. Lee, C.H.; Chen, Y.J.; Liao, C.H.; Popuri, R.S.; Tsai, S.L.; Hung, C.E. Selective leaching process for Neodymium recovery from scrap Nd-Fe-B magnet. *Metall. Mater. Trans. A Phys. Metall. Mater. Sci.* **2013**, *44*, 5825–5833. [CrossRef]
11. Rabatho, J.P.; Tongamp, W.; Takasaki, Y.; Haga, K.; Shibayama, A. Recovery of Nd and Dy from rare earth magnetic waste sludge by hydrometallurgical process. *J. Mater. Cycles Waste Manag.* **2013**, *15*, 171–178. [CrossRef]
12. Yingnakorn, T.; Laokhen, P.; Sriklang, L.; Patcharawit, T.; Khumkoa, S. Study on Recovery of Rare Earth Elements from NdFeB Magnet Scrap by Using Selective Leaching. *Mater. Sci. Forum* **2020**, *1009*, 149–154. [CrossRef]
13. Feng, L.Y.; Zhu, M.G.; Wei, L.; Dong, Z.; Feng, L.; Lang, C.; Wu, J.Y.; Yan, Q.; An, D. The Impact Induced Demagnetization Mechanism in NdFeB Permanent Magnets. *Chin. Phys. Lett.* **2013**, *30*, 097501.
14. Hoogerstraete, T.V.; Blanpain, B.; Gerven, T.V.; Binnemans, K. From NdFeB magnets towards the rare-earth oxides: A recycling process consuming only oxalic acid. *RSC Adv.* **2014**, *4*, 64099–64111. [CrossRef]
15. Riano, S.; Binnemans, K. Extraction and separation of neodymium and dysprosium from used NdFeB magnets: An application of ionic liquids in solvent extraction towards the recycling of magnets. *Green Chem.* **2015**, *17*, 2931–2942. [CrossRef]
16. Onal, M.A.R.; Borra, C.R.; Guo, M.; Blanpain, B.; Van Gerven, T. Recycling of NdFeB magnets using sulfation, selective roasting, and water leaching. *J. Sustain. Metall.* **2015**, *1*, 199–215. [CrossRef]

MDPI
St. Alban-Anlage 66
4052 Basel
Switzerland
Tel. +41 61 683 77 34
Fax +41 61 302 89 18
www.mdpi.com

Metals Editorial Office
E-mail: metals@mdpi.com
www.mdpi.com/journal/metals

www.ingramcontent.com/pod-product-compliance
Lightning Source LLC
LaVergne TN
LVHW080501180726
843515LV00016B/596